Succeed

Eureka Math®
Grade 4
Modules 1–4

Published by Great Minds®.

Copyright © 2018 Great Minds®.

Printed in the U.S.A.

This book may be purchased from the publisher at eureka-math.org.

10 9 8 7 6 5 4 3 2

ISBN 978-1-64054-090-3

G4-M1-M4-S-06.2018

Learn ◆ Practice ◆ Succeed

Eureka Math® student materials for *A Story of Units*® (K–5) are available in the *Learn, Practice, Succeed* trio. This series supports differentiation and remediation while keeping student materials organized and accessible. Educators will find that the *Learn, Practice,* and *Succeed* series also offers coherent—and therefore, more effective—resources for Response to Intervention (RTI), extra practice, and summer learning.

Learn

Eureka Math Learn serves as a student's in-class companion where they show their thinking, share what they know, and watch their knowledge build every day. *Learn* assembles the daily classwork—Application Problems, Exit Tickets, Problem Sets, templates—in an easily stored and navigated volume.

Practice

Each *Eureka Math* lesson begins with a series of energetic, joyous fluency activities, including those found in *Eureka Math Practice*. Students who are fluent in their math facts can master more material more deeply. With *Practice,* students build competence in newly acquired skills and reinforce previous learning in preparation for the next lesson.

Together, *Learn* and *Practice* provide all the print materials students will use for their core math instruction.

Succeed

Eureka Math Succeed enables students to work individually toward mastery. These additional problem sets align lesson by lesson with classroom instruction, making them ideal for use as homework or extra practice. Each problem set is accompanied by a Homework Helper, a set of worked examples that illustrate how to solve similar problems.

Teachers and tutors can use *Succeed* books from prior grade levels as curriculum-consistent tools for filling gaps in foundational knowledge. Students will thrive and progress more quickly as familiar models facilitate connections to their current grade-level content.

Students, families, and educators:

Thank you for being part of the *Eureka Math®* community, where we celebrate the joy, wonder, and thrill of mathematics.

Nothing beats the satisfaction of success—the more competent students become, the greater their motivation and engagement. The *Eureka Math Succeed* book provides the guidance and extra practice students need to shore up foundational knowledge and build mastery with new material.

What is in the Succeed *book?*

Eureka Math Succeed books deliver supported practice sets that parallel the lessons of *A Story of Units®*. Each *Succeed* lesson begins with a set of worked examples, called *Homework Helpers*, that illustrate the modeling and reasoning the curriculum uses to build understanding. Next, students receive scaffolded practice through a series of problems carefully sequenced to begin from a place of confidence and add incremental complexity.

How should Succeed *be used?*

The collection of *Succeed* books can be used as differentiated instruction, practice, homework, or intervention. When coupled with *Affirm®*, *Eureka Math*'s digital assessment system, *Succeed* lessons enable educators to give targeted practice and to assess student progress. *Succeed*'s perfect alignment with the mathematical models and language used across *A Story of Units* ensures that students feel the connections and relevance to their daily instruction, whether they are working on foundational skills or getting extra practice on the current topic.

Where can I learn more about Eureka Math *resources?*

The Great Minds® team is committed to supporting students, families, and educators with an ever-growing library of resources, available at eureka-math.org. The website also offers inspiring stories of success in the *Eureka Math* community. Share your insights and accomplishments with fellow users by becoming a *Eureka Math* Champion.

Best wishes for a year filled with Eureka moments!

Jill Diniz

Jill Diniz
Director of Mathematics
Great Minds

Contents

Module 1: Place Value, Rounding, and Algorithms for Addition and Subtraction

Module 2: Unit Conversions and Problem Solving with Metric Measurement

Module 3: Multi-Digit Multiplication and Division

Topic F: Reasoning with Divisibility

Topic G: Division of Thousands, Hundreds, Tens, and Ones

Topic H: Multiplication of Two-Digit by Two-Digit Numbers

Module 4: Angle Measure and Plane Figures

Topic A: Lines and Angles

Topic B: Angle Measurement

Topic C: Problem Solving with the Addition of Angle Measures

Topic D: Two-Dimensional Figures and Symmetry

Grade 4
Module 1

1. Label the place value charts. Fill in the blanks to make the following equations true. Draw disks in the place value chart to show how you got your answer, using arrows to show any regrouping.

10 × 3 ones = _____**30**_____ ones = _____**3 tens**_____

thousands	hundreds	tens	ones

10 × 3 ones is represented by drawing 3 disks in the ones column and then drawing 9 more ones for each disk. 10 × 3 ones is 30 ones.

thousands	hundreds	tens	ones

I draw an arrow to the tens column to show I am regrouping 10 ones as 1 ten. 30 ones is the same as 3 tens.

2. Complete the following statements using your knowledge of place value. Then, use pictures, numbers, or words to explain how you got your answer.

____60____ hundreds is the same as 6 thousands.

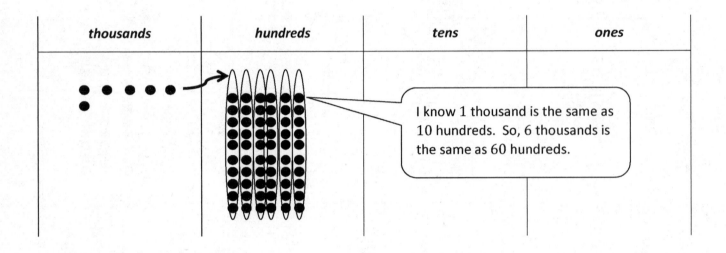

thousands	hundreds	tens	ones

I know 1 thousand is the same as 10 hundreds. So, 6 thousands is the same as 60 hundreds.

3. Gabby has 50 books in her room. Her mom has 10 times as many books in her office. How many books does Gabby's mom have? Use numbers or words to explain how you got your answer.

5 *tens* × 10 = 50 *tens*

Gabby's mom has 500 books in her office.

50 tens is the same as 5 hundreds. I can write my answer in standard form within a sentence to explain my answer.

Lesson 1:   Interpret a multiplication equation as a comparison.

EUREKA
MATH

Name _____   Date _____

1. Label the place value charts. Fill in the blanks to make the following equations true. Draw disks in the place value chart to show how you got your answer, using arrows to show any regrouping.

a. 10 × 4 ones = _____ ones = _____

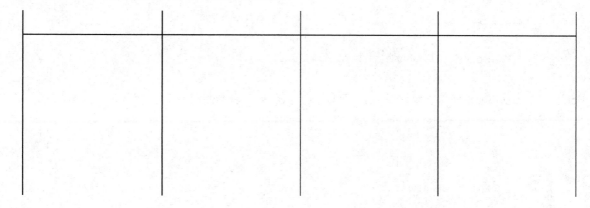

b. 10 × 2 tens =_____ tens = _____

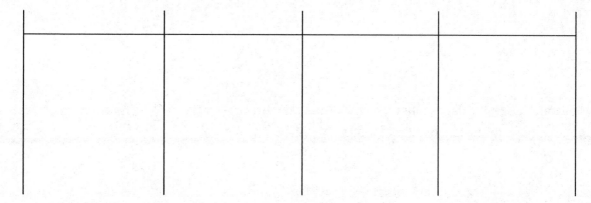

c. 5 hundreds × 10 = _____ hundreds = _____

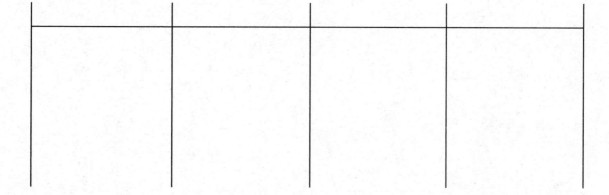

2. Complete the following statements using your knowledge of place value:

   a. 10 times as many as 1 hundred is _____ hundreds or _____ thousand.

   b. 10 times as many as _____ hundreds is 60 hundreds or _____ thousands.

   c. _____ as 8 hundreds is 8 thousands.

   d. _____ hundreds is the same as 4 thousands.

   Use pictures, numbers, or words to explain how you got your answer for Part (d).

3. Katrina has 60 GB of storage on her tablet. Katrina's father has 10 times as much storage on his computer. How much storage does Katrina's father have? Use numbers or words to explain how you got your answer.

Lesson 1:   Interpret a multiplication equation as a comparison.

EUREKA
MATH

4. Katrina saved $200 to purchase her tablet. Her father spent 10 times as much money to buy his new computer. How much did her father's computer cost? Use numbers or words to explain how you got your answer.

5. Fill in the blanks to make the statements true.

   a. 4 times as much as 3 is _____.

   b. 10 times as much as 9 is _____.

   c. 700 is 10 times as much as _____.

   d. 8,000 is _____ as 800.

6. Tomas's grandfather is 100 years old. Tomas's grandfather is 10 times as old as Tomas. How old is Tomas?

1. Label and represent the product or quotient by drawing disks on the place value chart.

   a.   10 × 3 thousands =   **30**   thousands =   **3 _ten thousands_**

millions	hundred thousands	ten thousands	thousands	hundreds	tens	ones

Just as in Lesson 1, I group each ten with a circle and draw an arrow to show I am regrouping 30 thousands as 3 ten thousands.

   b.   2 thousands ÷ 10 = __**20**__ hundreds ÷ 10 = __**2 hundreds**__

millions	hundred thousands	ten thousands	thousands	hundreds	tens	ones

I can't divide 2 thousands disks into equal groups of 10. So, I rename 2 thousands as 20 hundreds. Now, I can divide 20 hundreds into equal groups of 10.

EUREKA MATH

Lesson 2:   Recognize a digit represents 10 times the value of what it represents in the place to its right.

© 2018 Great Minds®. eureka-math.org

9

2. Solve for the expression by writing the solution in unit form and in standard form.

Expression	Unit Form	Standard Form
(3 tens 2 ones) × 10	**30 *tens* 20 *ones***	**320**

> I multiply each unit, the tens and the ones, by 10.

3. Solve.

840 matches are in 1 box. 10 times as many matches are in a package. How many matches in a package?

84 *tens* × 10 *is* 840 *tens* or 84 *hundreds*.

840 × 10 = 8,400

8,400 *matches are in a package*.

> I can use unit form to make the multiplication easier and to verify my answer in standard form.

Lesson 2: Recognize a digit represents 10 times the value of what it represents in the place to its right.

EUREKA MATH®

Name _____   Date _____

1. As you did during the lesson, label and represent the product or quotient by drawing disks on the place value chart.

    a.   10 × 4 thousands = _____ thousands = _____

    b.   4 thousands ÷ 10 = _____ hundreds ÷ 10 = _____

2. Solve for each expression by writing the solution in unit form and in standard form.

Expression	Unit Form	Standard Form
10 × 3 tens		
5 hundreds × 10		
9 ten thousands ÷ 10		
10 × 7 thousands		

   Lesson 2:   Recognize a digit represents 10 times the value of what it represents in the   11
                        place to its right.

© 2018 Great Minds®. eureka-math.org

3. Solve for each expression by writing the solution in unit form and in standard form.

Expression	Unit Form	Standard Form
(2 tens 1 one) × 10		
(5 hundreds 5 tens) × 10		
(2 thousands 7 tens) ÷ 10		
(4 ten thousands 8 hundreds) ÷ 10		

4. a. Emily collected $950 selling Girl Scout cookies all day Saturday. Emily's troop collected 10 times as much as she did. How much money did Emily's troop raise?

   b. On Saturday, Emily made 10 times as much as on Monday. How much money did Emily collect on Monday?

Lesson 2: Recognize a digit represents 10 times the value of what it represents in the place to its right.

© 2018 Great Minds®. eureka-math.org

EUREKA
MATH

1. Rewrite the following number, including commas where appropriate:

30030033003   __30,030,033,003__

> I use a comma after every 3 digits from the right to indicate the periods, or grouping of units—ones, thousands, millions, and billions.

2. Solve each expression. Record your answer in standard form.

> I can add 5 tens + 9 tens = 14 tens.

Expression	Standard Form
5 tens + 9 tens	140

> 14 tens is the same as 10 tens and 4 tens. I can bundle 10 tens to make 1 hundred. 14 tens is the same as 140.

3. Represent each addend with place value disks in the place value chart. Show the composition of larger units from 10 smaller units. Write the sum in standard form.

3 thousands + 14 hundreds = __4,400__

millions	hundred thousands	ten thousands	thousands	hundreds	tens	ones
			• • • ●	● ● ● ● ● ● ● ● ● ● ● ● ● ● ● ● ● ●		

> After drawing 3 thousands and 14 hundreds disks, I notice that 10 hundreds can be bundled as 1 thousand. Now, my picture shows 4 thousands 4 hundreds, or 4,400.

EUREKA
MATH®

Lesson 3:   Name numbers within 1 million by building understanding of the place value chart and placement of commas for naming base thousand units.

13

© 2018 Great Minds®. eureka-math.org

4. Use digits or disks on the place value chart to represent the following equations. Write the product in standard form.

(5 ten thousands 3 thousands) $\times$ 10 $=$ __530,000__

How many thousands are in your answer?   __530 *thousands*__

> The place value to the left represents 10 times as much, so I can draw an arrow and label it "$\times$ 10".

millions	hundred thousands	ten thousands	thousands	hundreds	tens	ones

$\times$ 10 $\times$ 10

> 3 ten thousands is 10 times more than 3 thousands. 5 hundred thousands is 10 times more than 5 ten thousands. So, (5 ten thousands 3 thousands) $\times$ 10 is 530,000.

Lesson 3:   Name numbers within 1 million by building understanding of the place value chart and placement of commas for naming base thousand units.

EUREKA MATH

Name _____   Date _____

1. Rewrite the following numbers including commas where appropriate:

    a. 4321   _____   b. 54321   _____

    c. 224466 _____   d. 2224466 _____

    e. 10010011001 _____

2. Solve each expression. Record your answer in standard form.

Expression	Standard Form
4 tens + 6 tens	
8 hundreds + 2 hundreds	
5 thousands + 7 thousands	

3. Represent each addend with place value disks in the place value chart. Show the composition of larger units from 10 smaller units. Write the sum in standard form.

    a.   2 thousands + 12 hundreds = _____

millions	hundred thousands	ten thousands	thousands	hundreds	tens	ones

Lesson 3:   Name numbers within 1 million by building understanding of the place value chart and placement of commas for naming base thousand units.

© 2018 Great Minds®. eureka-math.org

b.   14 ten thousands + 12 thousands = _____

millions	hundred thousands	ten thousands	thousands	hundreds	tens	ones

4. Use digits or disks on the place value chart to represent the following equations. Write the product in standard form.

a.   10 × 5 thousands = _____

How many thousands are in the answer? _____

millions	hundred thousands	ten thousands	thousands	hundreds	tens	ones

b.   (4 ten thousands 4 thousands) × 10 = _____

How many thousands are in the answer? _____

millions	hundred thousands	ten thousands	thousands	hundreds	tens	ones

Lesson 3:   Name numbers within 1 million by building understanding of the place value chart and placement of commas for naming base thousand units.

EUREKA
MATH®

c. (27 thousands 3 hundreds 5 ones) × 10 = _____

How many thousands are in your answer? _____

millions	hundred thousands	ten thousands	thousands	hundreds	tens	ones

5. A large grocery store received an order of 2 thousand apples. A neighboring school received an order of 20 boxes of apples with 100 apples in each. Use digits or disks on a place value chart to compare the number of apples received by the school and the number of apples received by the grocery store.

EUREKA MATH

Lesson 3:   Name numbers within 1 million by building understanding of the place value chart and placement of commas for naming base thousand units.

© 2018 Great Minds®. eureka-math.org

17

1.

   a.   On the place value chart below, label the units, and represent the number 43,082.

millions	hundred thousands	ten thousands	thousands	hundreds	tens	ones
		● ● ● ●	● ● ●		● ● ● ● ● ● ● ●	● ●

   b. Write the number in word form.   ***forty-three thousand, eighty-two***

> I read 43,082 to myself. I write the words that I say. I add commas to separate the periods of thousands and ones, just as I do when I write numerals.

   c. Write the number in expanded form. $40,000 + 3,000 + 80 + 2$

> I write the value of each digit in 43,082 as an addition expression. The 4 has a value of 4 ten thousands, which I write in standard form as 40,000. $43,082 = 40,000 + 3,000 + 80 + 2.$

2. Use pictures, numbers, and words to explain another way to say 39 hundred.

   Another way to say 39 hundred is 3 thousand, 9 hundred. I can write 3, 900, and I draw 39 hundreds disks as 3 thousands disks and 9 hundreds disks.

millions	hundred thousands	ten thousands	thousands	hundreds	tens	ones

I know 10 hundreds is the same as 1 thousand. I can bundle 30 hundreds to make 3 thousands.

Lesson 4:   Read and write multi-digit numbers using base ten numerals, number names, and expanded form.

© 2018 Great Minds®. eureka-math.org

EUREKA MATH

Name _____   Date _____

1. a.   On the place value chart below, label the units, and represent the number 50,679.

    b.   Write the number in word form.

    c.   Write the number in expanded form.

2. a.   On the place value chart below, label the units, and represent the number 506,709.

    b.   Write the number in word form.

    c.   Write the number in expanded form.

Lesson 4:   Read and write multi-digit numbers using base ten numerals, number   **21**
                   names, and expanded form.

© 2018 Great Minds®. eureka-math.org

3. Complete the following chart:

Standard Form	Word Form	Expanded Form
	five thousand, three hundred seventy	
		50,000 + 300 + 70 + 2
	thirty-nine thousand, seven hundred one	
309,017		
770,070		

4. Use pictures, numbers, and words to explain another way to say sixty-five hundred.

Lesson 4:   Read and write multi-digit numbers using base ten numerals, number names, and expanded form.

© 2018 Great Minds®. eureka-math.org

EUREKA
MATH®

1. Label the units in the place value chart. Draw place value disks to represent each number in the place value chart. Use <, >, or = to compare the two numbers. Write the correct symbol in the circle.

503,421 (>) 350,491

> I record the comparison symbol for *greater than*.

millions	hundred thousands	ten thousands	thousands	hundreds	tens	ones
	● ● ● ● ●		● ● ●	● ● ● ●	● ●	●
	● ● ●	● ● ● ● ●		● ● ● ●	● ● ● ● ● ● / ● ● ● ●	●

> I record the value of each digit using place value disks, placing 503,421 in the top half and 350,491 in the bottom half of the place value chart. I can clearly see and compare the unit with the greatest value—hundred thousands. 5 hundred thousands is greater than 3 hundred thousands. 503,421 is greater than 350,491.

2. Compare the two numbers by using the symbols <, >, or =. Write the correct symbol in the circle.

six hundred two thousand, four hundred seventy-three (<) 600,000 + 50,000 + 2,000 + 700 + 7

> It helps me to solve if I write both numbers in standard form.

 602,473 (<) 652,707

> Since the value of the largest unit is the same, I compare the next largest unit—the ten thousands. Zero ten thousands is less than five ten thousands. So, 602,473 is less than 652,707. I record the comparison symbol for *less than* to complete my answer.

3. Jill has $1,462, Adam has $1,509, Cristina has $1,712, and Robin has $1,467. Arrange the amounts of money in order from greatest to least. Then, name who has the most money.

thousands	hundreds	tens	ones
1	4	6	2
1	5	0	9
1	7	1	2
1	4	6	7

> Listing the amounts of money in a place value chart helps me to see the values in each unit.

$$\$1,712 > \$1,509 > \$1,467 > \$1,462$$

Cristina has the most money.

> I notice 1,462 and 1,467 both have 1 thousand, 4 hundreds, *and* 6 tens. So, I compare the ones. 7 ones is more than 2 ones. 1,467 is greater than 1,462.

Lesson 5:   Compare numbers based on meanings of the digits using <, >, or = to record the comparison.

© 2018 Great Minds®. eureka-math.org

Name _____   Date _____

1. Label the units in the place value chart. Draw place value disks to represent each number in the place value chart. Use <, >, or = to compare the two numbers. Write the correct symbol in the circle.

    a.   909,013 ◯ 90,013

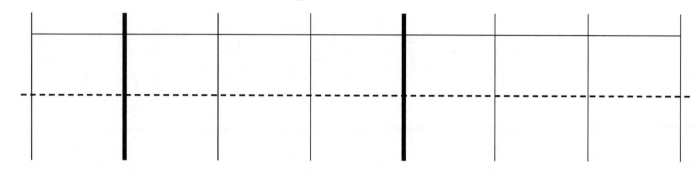

    b.   210,005 ◯ 220,005

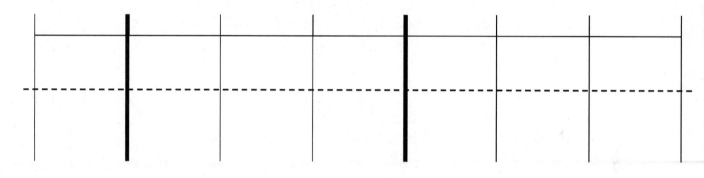

EUREKA MATH

Lesson 5:   Compare numbers based on meanings of the digits using >, <, or = to record the comparison.

© 2018 Great Minds®. eureka-math.org

25

2. Compare the two numbers by using the symbols <, >, and =. Write the correct symbol in the circle.

   a. 501,107 ◯ 89,171

   b. 300,000 + 50,000 + 1,000 + 800 ◯ six hundred five thousand, nine hundred eight

   c. 3 hundred thousands 3 thousands 8 hundreds 4 tens ◯ 303,840

   d. 5 hundreds 6 ten thousands 2 ones ◯ 3 ten thousands 5 hundreds 1 one

3. Use the information in the chart below to list the height, in feet, of each skyscraper from shortest to tallest. Then, name the tallest skyscraper.

Name of Skyscraper	Height of Skyscraper (ft)
Willis Tower	1,450 ft
One World Trade Center	1,776 ft
Taipei 101	1,670 ft
Petronas Towers	1,483 ft

Lesson 5: Compare numbers based on meanings of the digits using >, <, or = to record the comparison.

© 2018 Great Minds®. eureka-math.org

EUREKA MATH

4. Arrange these numbers from least to greatest: 7,550   5,070   750   5,007   7,505

5. Arrange these numbers from greatest to least: 426,000   406,200   640,020   46,600

6. The areas of the 50 states can be measured in square miles.

   California is 158,648 square miles. Nevada is 110,567 square miles. Arizona is 114,007 square miles.
   Texas is 266,874 square miles. Montana is 147,047 square miles, and Alaska is 587,878 square miles.

   Arrange the states in order from least area to greatest area.

Lesson 5:   Compare numbers based on meanings of the digits using >, <, or = to
record the comparison.

© 2018 Great Minds®. eureka-math.org

27

1. Round to the nearest thousand. Use the number line to model your thinking.

a. $3,941 \approx$ **4,000**   b. $53,269 \approx$ **53,000**

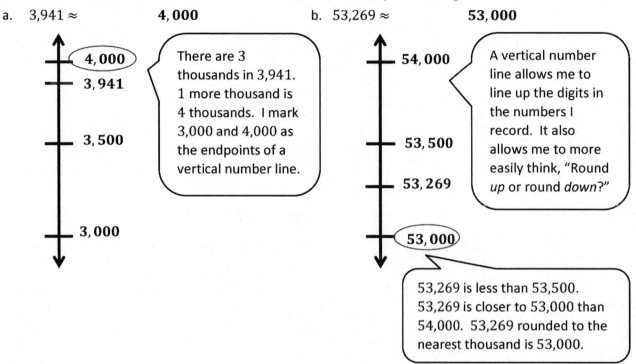

There are 3 thousands in 3,941. 1 more thousand is 4 thousands. I mark 3,000 and 4,000 as the endpoints of a vertical number line.

A vertical number line allows me to line up the digits in the numbers I record. It also allows me to more easily think, "Round up or round *down?*"

53,269 is less than 53,500. 53,269 is closer to 53,000 than 54,000. 53,269 rounded to the nearest thousand is 53,000.

2. In 2013, the family vacation cost $3,809. In 2014, the family vacation cost $4,699. The family budgeted about $4,000 for each vacation. In which year did the family stay closer to their budget? Round to the nearest thousand. Use what you know about place value to explain your answer.

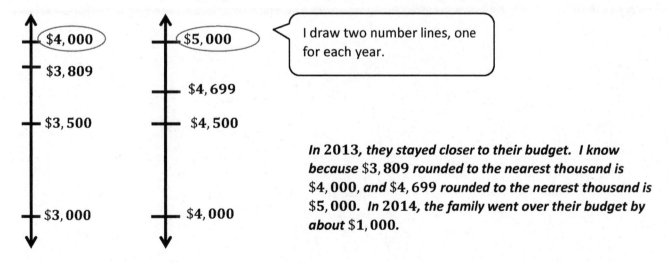

I draw two number lines, one for each year.

In 2013, they stayed closer to their budget. I know because $3,809 rounded to the nearest thousand is $4,000, and $4,699 rounded to the nearest thousand is $5,000. In 2014, the family went over their budget by about $1,000.

Lesson 7:   Round multi-digit numbers to the thousands place using the vertical number line.

© 2018 Great Minds®. eureka-math.org

35

Name _____   Date _____

1. Round to the nearest thousand. Use the number line to model your thinking.

     a.   5,900 ≈ _____

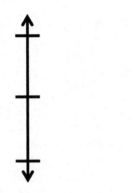

     b.   4,180 ≈ _____

     c.   32,879 ≈ _____

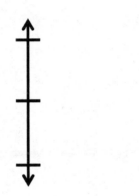

     d.   78,600 ≈ _____

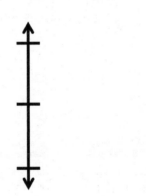

     e.   251,031 ≈ _____

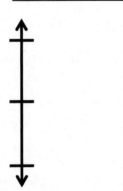

     f.   699,900 ≈ _____

2. Steven put together 981 pieces of a puzzle. About how many pieces did he put together? Round to the nearest thousand. Use what you know about place value to explain your answer.

3. Louise's family went on vacation to Disney World. Their vacation cost $5,990. Sophia's family went on vacation to Niagara Falls. Their vacation cost $4,720. Both families budgeted about $5,000 for their vacation. Whose family stayed closer to the budget? Round to the nearest thousand. Use what you know about place value to explain your answer.

4. Marsha's brother wanted help with the first question on his homework. The question asked the students to round 128,902 to the nearest thousand and then to explain the answer. Marsha's brother thought that the answer was 128,000. Was his answer correct? How do you know? Use pictures, numbers, or words to explain.

Lesson 7: Round multi-digit numbers to the thousands place using the vertical number line.

© 2018 Great Minds®. eureka-math.org

EUREKA
MATH

1. Complete each statement by rounding the number to the given place value. Use the number line to show your work.

   a. 41,899 rounded to the nearest ten thousand is   **40,000**

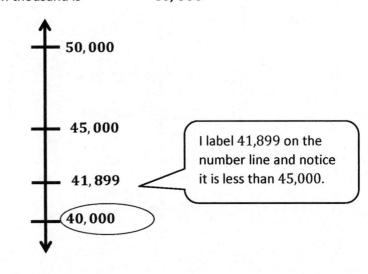

I ask myself, "How many ten thousands in 41,899? What is 1 more ten thousand?"

I label 41,899 on the number line and notice it is less than 45,000.

50,000

45,000

41,899

40,000

   b. 267,072 rounded to the nearest hundred thousand is   **300,000**

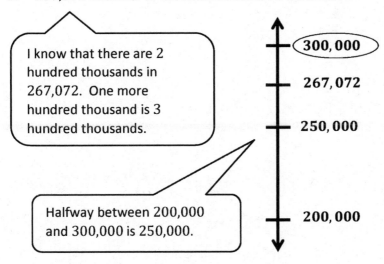

I know that there are 2 hundred thousands in 267,072. One more hundred thousand is 3 hundred thousands.

Halfway between 200,000 and 300,000 is 250,000.

300,000

267,072

250,000

200,000

2. 982,510 books were downloaded in one year. Round this number to the nearest hundred thousand to estimate how many books were downloaded in one year. Use a number line to show your work.

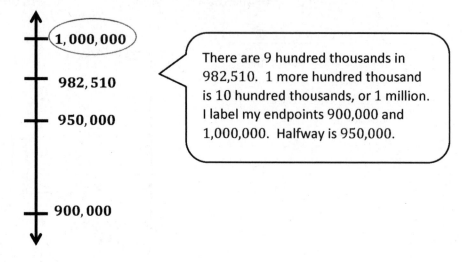

There are 9 hundred thousands in 982,510. 1 more hundred thousand is 10 hundred thousands, or 1 million. I label my endpoints 900,000 and 1,000,000. Halfway is 950,000.

About 1 million books were downloaded in one year.

3. Estimate the difference by rounding each number to the given place value.

$$519,240 - 339,705$$

a. Round to the nearest hundred thousand.

$$500,000 - 300,000 = 200,000$$

b. Round to the nearest ten thousand.

$$520,000 - 340,000 = 180,000$$

Thinking in unit language makes this subtraction easy: 520 thousands minus 340 thousands equals 180 thousands.

EUREKA
MATH®

Name _____   Date _____

Complete each statement by rounding the number to the given place value. Use the number line to show your work.

1. a. 67,000 rounded to the nearest ten thousand is _____.

   b. 51,988 rounded to the nearest ten thousand is _____.

   c. 105,159 rounded to the nearest ten thousand is _____.

2. a. 867,000 rounded to the nearest hundred thousand is _____.

   b. 767,074 rounded to the nearest hundred thousand is _____.

   c. 629,999 rounded to the nearest hundred thousand is_____.

EUREKA MATH®

Lesson 8:   Round multi-digit numbers to any place using the vertical number line.

41

© 2018 Great Minds®. eureka-math.org

3. 491,852 people went to the water park in the month of July. Round this number to the nearest hundred thousand to estimate how many people went to the park. Use a number line to show your work.

4. This number was rounded to the nearest hundred thousand. List the possible digits that could go in the ten thousands place to make this statement correct. Use a number line to show your work.

$$1\_9{,}644 \approx 100{,}000$$

5. Estimate the sum by rounding each number to the given place value.

$$164{,}215 + 216{,}088$$

a. Round to the nearest ten thousand.

b. Round to the nearest hundred thousand.

EUREKA
MATH

1. Round to the nearest thousand.

   a. $7,598 \approx$ _____ **8,000** _____

   > I remember from Lesson 7 how to round to the nearest thousand.

   b. $301,409 \approx$ _____ **301,000** _____

   c. Explain how you found your answer for Part (b).

   There are 301 thousands in 301, 409. One more thousand is 302 thousands. Halfway between 301 thousands and 302 thousands is 301 thousands 5 hundreds. 301, 409 is less than 301, 500. Therefore, 301, 409 rounded to the nearest thousand is 301, 000.

2. Round to the nearest ten thousand.

   a. $73,999 \approx$ _____ **70,000** _____

   > I may need to draw a number line to verify my answer.

   b. $65,002 \approx$ _____ **70,000** _____

   c. Explain why the two problems have the same answer. Write another number that has the same answer when rounded to the nearest ten thousand.

   Any number equal to or greater than 65, 000 and less than 75, 000 will round to 70, 000 when rounded to the nearest ten thousand. 65, 002 is greater than 65, 000 and 73, 999 is less than 75, 000. Another number that would round to 70, 000 is 68, 234.

EUREKA MATH

Lesson 9:   Use place value understanding to round multi-digit numbers to any place value.

43

© 2018 Great Minds®. eureka-math.org

Solve the following problems using pictures, numbers, or words.

3. About 700,000 people make up the population of Americatown. If the population was rounded to the nearest hundred thousand, what could be the greatest and least number of people who make up the population of Americatown?

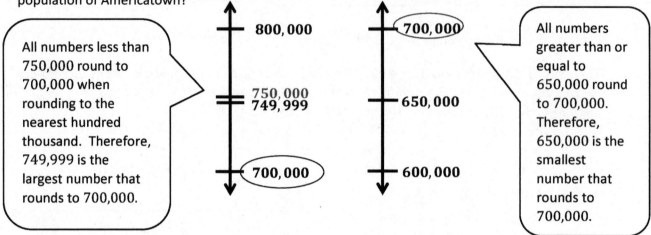

All numbers less than 750,000 round to 700,000 when rounding to the nearest hundred thousand. Therefore, 749,999 is the largest number that rounds to 700,000.

All numbers greater than or equal to 650,000 round to 700,000. Therefore, 650,000 is the smallest number that rounds to 700,000.

The greatest number of people that could make up the population is 749, 999. *I know because it is* 1 *fewer than* 750, 000. *The least number of people that could make up the population is* 650, 000.

Lesson 9: Use place value understanding to round multi-digit numbers to any place value.

© 2018 Great Minds®. eureka-math.org

EUREKA
MATH

Name _____ Date _____

1. Round to the nearest thousand.

   a. 6,842 ≈ _____   b. 2,722 ≈ _____

   c. 16,051 ≈ _____   d. 706,421 ≈ _____

   e. Explain how you found your answer for Part (d).

2. Round to the nearest ten thousand.

   a. 88,999 ≈ _____   b. 85,001 ≈ _____

   c. 789,091 ≈ _____   d. 905,154 ≈ _____

   e. Explain why two problems have the same answer. Write another number that has the same answer when rounded to the nearest ten thousand.

3. Round to the nearest hundred thousand.

   a. 89,659 ≈ _____   b. 751,447 ≈ _____

   c. 617,889 ≈ _____   d. 817,245 ≈ _____

   e. Explain why two problems have the same answer. Write another number that has the same answer when rounded to the nearest hundred thousand.

EUREKA MATH

Lesson 9:   Use place value understanding to round multi-digit numbers to any place value.

© 2018 Great Minds®. eureka-math.org

45

4. Solve the following problems using pictures, numbers, or words.

   a. At President Obama's inauguration in 2013, the newspaper headlines stated there were about 800,000 people in attendance. If the newspaper rounded to the nearest hundred thousand, what is the largest number and smallest number of people who could have been there?

   b. At President Bush's inauguration in 2005, the newspaper headlines stated there were about 400,000 people in attendance. If the newspaper rounded to the nearest ten thousand, what is the largest number and smallest number of people who could have been there?

   c. At President Lincoln's inauguration in 1861, the newspaper headlines stated there were about 30,000 people in attendance. If the newspaper rounded to the nearest thousand, what is the largest number and smallest number of people who could have been there?

Lesson 9:   Use place value understanding to round multi-digit numbers to any place value.

© 2018 Great Minds®. eureka-math.org

1. Round 745, 001 to the nearest

    > I remember from Lesson 7 to ask myself, "Between what two thousands is 745,001?" I try to picture the number line in my head.

    a. thousand: **745,000**

    b. ten thousand: **750,000**

    > I remember from Lesson 8 to find how many ten thousands and how many hundred thousands are in 745,001. Then, add one more of that unit to find the endpoints.

    c. hundred thousand: **700,000**

Solve the following problem using pictures, numbers, or words.

2. 37,248 people subscribe to the delivery of a local newspaper. To decide about how many papers to print, what place value should 37,248 be rounded to so each person receives a copy? Explain.

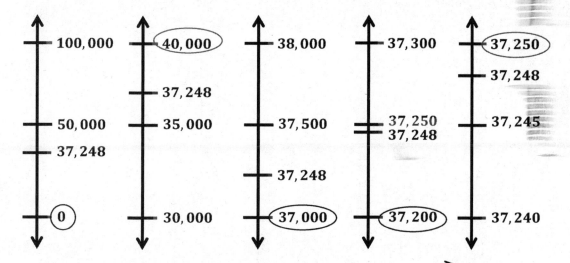

37, 248 should be rounded to the nearest ten thousand or the nearest ten. Extra papers will be printed, but if I round to the nearest hundred thousand, thousand, or hundred, there won't be enough papers printed.

> Drawing number lines helps to prove my written answer.

Lesson 10:   Use place value understanding to round multi-digit numbers to any place value using real world applications.   47

© 2018 Great Minds®. eureka-math.org

Name _____   Date _____

1. Round 845,001 to the nearest

    a. thousand: _____.

    b. ten thousand: _____.

    c. hundred thousand: _____.

2. Complete each statement by rounding the number to the given place value.

    a. 783 rounded to the nearest hundred is _____.

    b. 12,781 rounded to the nearest hundred is _____.

    c. 951,194 rounded to the nearest hundred is _____.

    d. 1,258 rounded to the nearest thousand is _____.

    e. 65,124 rounded to the nearest thousand is _____.

    f. 99,451 rounded to the nearest thousand is _____.

    g. 60,488 rounded to the nearest ten thousand is _____.

    h. 80,801 rounded to the nearest ten thousand is _____.

    i. 897,100 rounded to the nearest ten thousand is _____.

    j. 880,005 rounded to the nearest hundred thousand is _____.

    k. 545,999 rounded to the nearest hundred thousand is _____.

    l. 689,114 rounded to the nearest hundred thousand is _____.

Lesson 10:   Use place value understanding to round multi-digit numbers to any
place value using real world applications.

© 2018 Great Minds®. eureka-math.org

49

3. Solve the following problems using pictures, numbers, or words.

   a. In the 2011 New York City Marathon, 29,867 men finished the race, and 16,928 women finished the race. Each finisher was given a t-shirt. About how many men's shirts were given away? About how many women's shirts were given away? Explain how you found your answers.

   b. In the 2010 New York City Marathon, 42,429 people finished the race and received a medal. Before the race, the medals had to be ordered. If you were the person in charge of ordering the medals and estimated how many to order by rounding, would you have ordered enough medals? Explain your thinking.

   c. In 2010, 28,357 of the finishers were men, and 14,072 of the finishers were women. About how many more men finished the race than women? To determine your answer, did you round to the nearest ten thousand or thousand? Explain.

Lesson 10:   Use place value understanding to round multi-digit numbers to any place value using real world applications.

EUREKA MATH

> Using an algorithm means that the steps repeat themselves unit by unit. It can be an efficient way to solve a problem.

1. Solve the addition problems using the standard algorithm.

a.
```
   5,  1  2  2
+  2,  4  5  7
────────────────
   7,  5  7  9
```

b.
```
   5,  1  2  4
+  2,  4  5  7
        1
────────────────
   7,  5  8  1
```

c. $38,192 + 6,387 + 241,458$

```
    3  8,  1  9  2
        6,  3  8  7
+   2  4  1,  4  5  8
        1  1  2  1
─────────────────────
    2  8  6,  0  3  7
```

> No regroupings here! I just add like units. 2 ones plus 7 ones is 9 ones. I put the 9 in the ones column as part of the sum. Then, I continue to add the number of units of tens, the hundreds, and the thousands.

> I have to regroup ones. 4 ones + 7 ones = 11 ones. 11 ones equals 1 ten 1 one. I record 1 ten in the tens place on the line. I record 1 one in the ones column as part of the sum.

> The order of the addends doesn't matter as long as like units are lined up.

> I add tens. 2 tens + 5 tens + 1 ten = 8 tens. I record 8 tens in the tens column as part of the sum.

2. Draw a tape diagram to represent the problem. Use numbers to solve, and write your answer as a statement.

In July, the ice cream stand sold some ice cream cones. 3,907 were vanilla. 2,568 were not vanilla. How many cones did they sell in July?

> I can draw a tape diagram. I know the two parts, but I don't know the whole. I can label the unknown with a variable, C.

> I write an equation. Then, I solve to find the total. I write a statement to tell my answer.

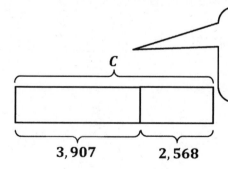

$$3,907 + 2,568 = C$$

```
   3,  9  0  7
+  2,  5  6  8
      1     1
─────────────────
   6,  4  7  5
```

The ice cream stand sold 6,475 cones in July.

EUREKA MATH   Lesson 11:   Use place value understanding to fluently add multi-digit whole numbers using the standard addition algorithm, and apply the algorithm to solve word problems using tape diagrams.   51

© 2018 Great Minds®. eureka-math.org

Name _____   Date _____

1. Solve the addition problems below using the standard algorithm.

a.   7,909
     + 1,044

b.   27,909
     +   9,740

c.   827,909
     +   42,989

d.   289,205
     +   11,845

e.   547,982
     + 114,849

f.   258,983
     + 121,897

g.   83,906
     +   35,808

h.   289,999
     +   91,849

i.   754,900
     + 245,100

EUREKA
MATH

Lesson 11:   Use place value understanding to fluently add multi-digit whole
              numbers using the standard addition algorithm, and apply the
              algorithm to solve word problems using tape diagrams.

© 2018 Great Minds®. eureka-math.org

53

Draw a tape diagram to represent each problem. Use numbers to solve, and write your answer as a statement.

2. At the zoo, Brooke learned that one of the rhinos weighs 4,897 pounds, one of the giraffes weighs 2,667 pounds, one of the African elephants weighs 12,456 pounds, and one of the Komodo dragons weighs 123 pounds.

   a. What is the combined weight of the zoo's African elephant and the giraffe?

   b. What is the combined weight of the zoo's African elephant and the rhino?

   c. What is the combined weight of the zoo's African elephant, the rhino, and the giraffe?

   d. What is the combined weight of the zoo's Komodo dragon and the rhino?

Lesson 11:   Use place value understanding to fluently add multi-digit whole numbers using the standard addition algorithm, and apply the algorithm to solve word problems using tape diagrams.

EUREKA
MATH

Estimate and then solve. Model the problem with a tape diagram. Explain if your answer is reasonable.

1. There were 4,806 more visitors to the zoo in the month of July than in the month of June. June had 6,782 visitors. How many visitors did the zoo have during both months?

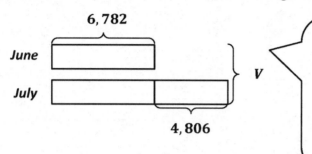

Since the problem states the relationship between June and July, I can draw two tapes. I make July's tape longer because there were more visitors in July. I partition July's tape into two parts: one part for the number of people in June and the other part for 4,806 more visitors.

a. About how many visitors did the zoo have during June and July?

$$7,000 + 7,000 + 5,000 = 19,000$$

The zoo had about 19,000 visitors during June and July.

To estimate the total, I round each number to the nearest thousand and add those numbers together.

b. Exactly how many visitors did the zoo have during June and July?

$$
\begin{array}{r}
6,782 \\
6,782 \\
+\ 4,806 \\
\hline
\scriptstyle 2\ \ 1\ \ 1 \\
\hline
18,370
\end{array}
$$

When I look at my tape diagram, I see that I don't have to solve for July to find the total. This saves me a step.

The zoo had exactly 18,370 visitors during June and July.

c. Is your answer reasonable? Compare your estimate to the answer. Write a sentence to explain your reasoning.

Sample Response: My answer is reasonable because my estimate of 19,000 is only about 600 more than the actual answer of 18,370. My estimate is greater than the actual answer because I rounded each addend up to the next thousand.

EUREKA MATH

Lesson 12:   Solve multi-step word problems using the standard addition algorithm modeled with tape diagrams, and assess the reasonableness of answers using rounding.

© 2018 Great Minds®. eureka-math.org

55

2. Emma's class spent four months collecting pennies.

a. During Month 3, the class collected 1,211 more pennies than they did during Month 2. Find the total number of pennies collected in four months.

Month	Pennies Collected
1	4,987
2	8,709
3	
4	8,192

Month 1 | 4,987 |

Month 2 | 8,709 | 1,211 |

Month 3 | 8,709 |

Month 4 | 8,192 |

P

> I draw four tapes to represent each month. Now, I can see how many pennies were collected in Month 3.

$$5,000 + 9,000 + 9,000 + 1,000 + 8,000 = 32,000$$

$$
\begin{array}{r}
4,\ 9\ \ 8\ \ 7 \\
8,\ 7\ \ 0\ \ 9 \\
8,\ 7\ \ 0\ \ 9 \\
1,\ 2\ \ 1\ \ 1 \\
+\ 8,\ 1\ \ 9\ \ 2 \\
\hline
3\ 1,\ 8\ \ 0\ \ 8 \\
\end{array}
$$

> I add in unit form: 5 thousands + 9 thousands + 9 thousands + 1 thousand + 8 thousands = 32 thousands. 32 thousand is an estimate of the total number of pennies collected in four months.

> To find the total pennies collected in the four months, I could solve for Month 3 and then add all of the months together to solve for P. Instead, I just add the value of each of the tapes together. The tape diagram shows me how to solve this in one step, not two.

The total number of pennies collected in four months was $31,808$.

b. Is your answer reasonable? Explain.

Sample Response: My answer is reasonable. $31,808$ is only about 200 less than the estimate of $32,000$.

Lesson 12:   Solve multi-step word problems using the standard addition algorithm modeled with tape diagrams, and assess the reasonableness of answers using rounding.
© 2018 Great Minds®. eureka-math.org

EUREKA MATH®

Name _____   Date _____

Estimate and then solve each problem. Model the problem with a tape diagram. Explain if your answer is reasonable.

1. There were 3,905 more hits on the school's website in January than February. February had 9,854 hits. How many hits did the school's website have during both months?

   a. About how many hits did the website have during January and February?

   b. Exactly how many hits did the website have during January and February?

   c. Is your answer reasonable? Compare your estimate from (a) to your answer from (b). Write a sentence to explain your reasoning.

EUREKA MATH®

Lesson 12:   Solve multi-step word problems using the standard addition algorithm modeled with tape diagrams, and assess the reasonableness of answers using rounding.

© 2018 Great Minds®. eureka-math.org

57

2. On Sunday, 77,098 fans attended a New York Jets game. The same day, 3,397 more fans attended a New York Giants game than attended the Jets game. Altogether, how many fans attended the games?

   a. What was the actual number of fans who attended the games?

   b. Is your answer reasonable? Round each number to the nearest thousand to find an estimate of how many fans attended the games.

Lesson 12:   Solve multi-step word problems using the standard addition algorithm modeled with tape diagrams, and assess the reasonableness of answers using rounding.
© 2018 Great Minds®. eureka-math.org

EUREKA MATH

3. Last year on Ted's farm, his four cows produced the following number of liters of milk:

Cow	Liters of Milk Produced
Daisy	5,098
Betsy	
Mary	9,980
Buttercup	7,087

a. Betsy produced 986 more liters of milk than Buttercup. How many liters of milk did all 4 cows produce?

b. Is your answer reasonable? Explain.

EUREKA
MATH®

Lesson 12:   Solve multi-step word problems using the standard addition algorithm
modeled with tape diagrams, and assess the reasonableness of
answers using rounding.
© 2018 Great Minds®. eureka-math.org

59

> I don't have enough tens to subtract 5 tens from 3 tens. I decompose 1 hundred for 10 tens.

1. Use the standard algorithm to solve the following subtraction problems.

a.
```
    6,  5  6  7
  - 1,  4  5  7
  ─────────────
    5,  1  1  0
```

> I look across the top number to see if I can subtract. I have enough units, so no regroupings! I just subtract like units. 7 ones minus 7 ones is 0 ones. I continue to subtract the number of units of tens, hundreds, and thousands.

b.
```
         4  13
    6,  5̶  3̶  7
  - 2,  4  5  7
  ─────────────
    4,  0  8  0
```

> Now, I have 4 hundreds. I show this by crossing off the 5 and writing a 4 in the hundreds place instead. 10 tens + 3 tens = 13 tens. I show this by crossing off the 3 tens and writing 13 in the tens place instead.

c. $3,532 - 921$

```
    2  15
    3̶, 5̶  3  2
  -     9  2  1
  ─────────────
    2,  6  1  1
```

> Just like in Lesson 11, I write the problem in vertical form, being sure to line up the units.

2. What number must be added to 23,165 to result in a sum of 46,884?

> To solve a word problem, I use RDW: Read, Draw, Write. I read the problem. I draw a picture, like a tape diagram, and I write my answer as an equation and a statement.

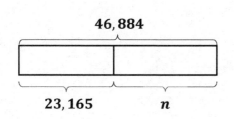

46,884

23,165 n

$$23,165 + n = 46,884$$

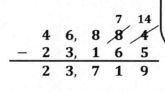

```
         7  14
    4  6, 8  8̶  4̶
  - 2  3, 1  6  5
  ────────────────
    2  3, 7  1  9
```

23,719 must be added to 23,165.

EUREKA MATH

Lesson 13:   Use place value understanding to decompose to smaller units once using the standard subtraction algorithm, and apply the algorithm to solve word problems using tape diagrams.

© 2018 Great Minds®. eureka-math.org

61

Draw a tape diagram to model the problem. Use numbers to solve, and write your answer as a statement. Check your answer.

3. Mr. Swanson drove his car 5,654 miles. Mrs. Swanson drove her car some miles, too. If they drove 11,965 miles combined, how many miles did Mrs. Swanson drive?

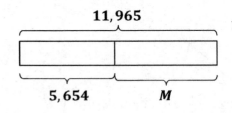

$11,965 - 5,654 = M$

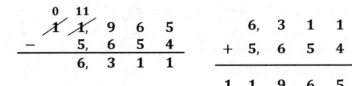

Mrs. Swanson drove 6,311 miles.

To check my answer, I add the difference to the known part. It equals the whole, so I subtracted correctly.

Lesson 13: Use place value understanding to decompose to smaller units once using the standard subtraction algorithm, and apply the algorithm to solve word problems using tape diagrams.
© 2018 Great Minds®. eureka-math.org

EUREKA MATH

Name _____ Date _____

1. Use the standard algorithm to solve the following subtraction problems.

a.
```
    2,431
  -   341
```

b.
```
   422,431
  -  14,321
```

c.
```
   422,431
  -  92,420
```

d.
```
   422,431
  - 392,420
```

e.
```
   982,430
  -  92,300
```

f.
```
   243,089
  - 137,079
```

g. 2,431 – 920 =

h. 892,431 – 520,800 =

2. What number must be added to 14,056 to result in a sum of 38,773?

EUREKA
MATH

Lesson 13:   Use place value understanding to decompose to smaller units once
using the standard subtraction algorithm, and apply the algorithm to
solve word problems using tape diagrams.

© 2018 Great Minds®. eureka-math.org

63

Draw a tape diagram to model each problem. Use numbers to solve, and write your answers as a statement. Check your answers.

3. An elementary school collected 1,705 bottles for a recycling program. A high school also collected some bottles. Both schools collected 3,627 bottles combined. How many bottles did the high school collect?

4. A computer shop sold $356,291 worth of computers and accessories. It sold $43,720 worth of accessories. How much did the computer shop sell in computers?

Lesson 13: Use place value understanding to decompose to smaller units once using the standard subtraction algorithm, and apply the algorithm to solve word problems using tape diagrams.

© 2018 Great Minds®. eureka-math.org

EUREKA
MATH

5. The population of a city is 538,381. In that population, 148,170 are children.

   a. How many adults live in the city?

   b. 186,101 of the adults are males. How many adults are female?

1. Use the standard algorithm to solve the following subtraction problems.

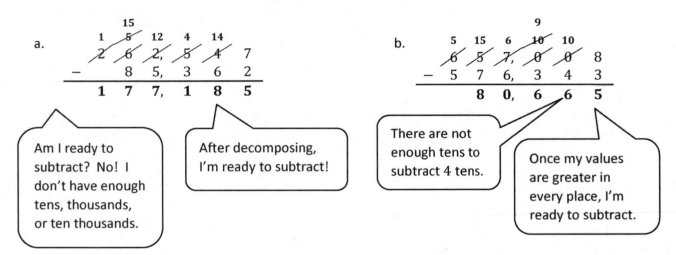

a.
```
        15
    1   5  12  4  14
    2   6  2,  5  4   7
  −     8  5,  3  6   2
  ─────────────────────
    1   7  7,  1  8   5
```

Am I ready to subtract? No! I don't have enough tens, thousands, or ten thousands.

After decomposing, I'm ready to subtract!

b.
```
                9
    5   15  6  10  10
    6   5  7,  0   0   8
  −     5  7,  6   3   4   3
  ─────────────────────────
        8  0,  6   6   5
```

There are not enough tens to subtract 4 tens.

Once my values are greater in every place, I'm ready to subtract.

Draw a tape diagram to represent the following problem. Use numbers to solve, and write your answer as a statement. Check your answer.

2. Stella had 542,000 visits to her website. Raquel had 231,348 visits to her website. How many more visits did Stella have than Raquel?

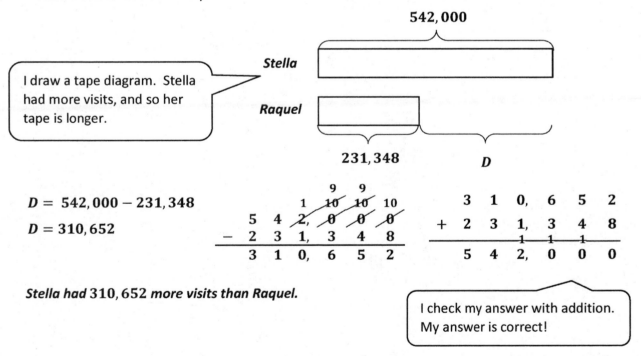

I draw a tape diagram. Stella had more visits, and so her tape is longer.

$D = 542,000 - 231,348$

$D = 310,652$

```
                9   9
            1  10  10  10
    5   4   2,  0   0   0
  − 2   3   1,  3   4   8
  ─────────────────────────
    3   1   0,  6   5   2
```

```
    3   1  0,  6   5   2
  + 2   3  1,  3   4   8
          1   1   1
  ─────────────────────────
    5   4  2,  0   0   0
```

I check my answer with addition. My answer is correct!

Stella had 310,652 more visits than Raquel.

EUREKA MATH

Lesson 14:   Use place value understanding to decompose to smaller units up to three times using the standard subtraction algorithm, and apply the algorithm to solve word problems using tape diagrams.

© 2018 Great Minds®. eureka-math.org

67

Name _____   Date _____

1. Use the standard algorithm to solve the following subtraction problems.

a.   71,989
    − 21,492

b.   371,989
    − 96,492

c.   371,089
    − 25,192

d.   879,989
    − 721,492

e.   879,009
    − 788,492

f.   879,989
    − 21,070

g.   879,000
    − 21,989

h.   279,389
    − 191,492

i.   500,989
    − 242,000

EUREKA MATH

Lesson 14:   Use place value understanding to decompose to smaller units up to three times using the standard subtraction algorithm, and apply the algorithm to solve word problems using tape diagrams.

© 2018 Great Minds®. eureka-math.org

69

Draw a tape diagram to represent each problem. Use numbers to solve, and write your answer as a statement. Check your answers.

2. Jason ordered 239,021 pounds of flour to be used in his 25 bakeries. The company delivering the flour showed up with 451,202 pounds. How many extra pounds of flour were delivered?

3. In May, the New York Public Library had 124,061 books checked out. Of those books, 31,117 were mystery books. How many of the books checked out were not mystery books?

4. A Class A dump truck can haul 239,000 pounds of dirt. A Class C dump truck can haul 600,200 pounds of dirt. How many more pounds can a Class C truck haul than a Class A truck?

Lesson 14: Use place value understanding to decompose to smaller units up to three times using the standard subtraction algorithm, and apply the algorithm to solve word problems using tape diagrams.

© 2018 Great Minds®. eureka-math.org

EUREKA
MATH

Use the standard subtraction algorithm to solve the problem below.

1.

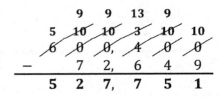

```
  6  0  0,  4  0  0
−     7  2,  6  4  9
```

> I am not ready to subtract. I must regroup.

Sample Student A Response:

```
         9   9  13   9
     5  10  10   3  10  10
     6   0   0,  4   0   0
  −      7   2,  6   4   9
     5   2   7,  7   5   1
```

> I work unit by unit, starting with the ones. I can rename 4 hundreds as 3 hundreds 10 tens. Then, I rename 10 tens as 9 tens 10 ones. I'll continue to decompose until I am ready to subtract.

Sample Student B Response:

```
                13
     5   9   9   3   9  10
     6   0   0,  4   0   0
  −      7   2,  6   4   9
     5   2   7,  7   5   1
```

> I need more ones. I unbundle 40 tens as 39 tens 10 ones.

> I need more than 3 hundreds to subtract 6 hundreds. I can rename the 600 thousands as 599 thousands 10 hundreds. 10 hundreds plus 3 hundreds is 13 hundreds.

EUREKA MATH

Lesson 15: Use place value understanding to fluently decompose to smaller units multiple times in any place using the standard subtraction algorithm, and apply the algorithm to solve word problems using tape diagrams.

© 2018 Great Minds®. eureka-math.org

71

Use a tape diagram and the standard algorithm to solve the problem below. Check your answer.

2. The cost of the Johnston's new home was \$200,000. They paid for most of it and now owe \$33,562. How much have they already paid?

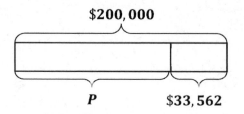

$$\$200,000 - \$33,562 = P$$

Sample Student A Response:

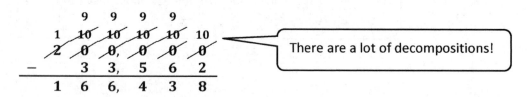

> There are a lot of decompositions!

Sample Student B Response:

> I rename 20,000 tens as 19,999 tens 10 ones.

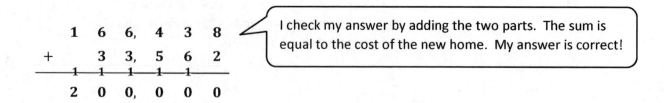

> I check my answer by adding the two parts. The sum is equal to the cost of the new home. My answer is correct!

The Johnstons have already paid $166,438.

EUREKA
MATH®

Name _____   Date _____

1. Use the standard subtraction algorithm to solve the problems below.

a.
```
   9,656
-    838
```

b.
```
  59,656
-  5,880
```

c.
```
  759,656
- 579,989
```

d.
```
  294,150
- 166,370
```

e.
```
  294,150
- 239,089
```

f.
```
  294,150
-  96,400
```

g.
```
  800,500
-  79,989
```

h.
```
  800,500
-  45,500
```

i.
```
  800,500
- 276,664
```

Use tape diagrams and the standard algorithm to solve the problems below. Check your answers.

2. A fishing boat was out to sea for 6 months and traveled a total of 8,578 miles. In the first month, the boat traveled 659 miles. How many miles did the fishing boat travel during the remaining 5 months?

EUREKA MATH

Lesson 15: Use place value understanding to fluently decompose to smaller units multiple times in any place using the standard subtraction algorithm, and apply the algorithm to solve word problems using tape diagrams.

© 2018 Great Minds®. eureka-math.org

73

3. A national monument had 160,747 visitors during the first week of September. A total of 759,656 people visited the monument in September. How many people visited the monument in September after the first week?

4. Shadow Software Company earned a total of $800,000 selling programs during the year 2012. $125,300 of that amount was used to pay expenses of the company. How much profit did Shadow Software Company make in the year 2012?

5. At the local aquarium, Bubba the Seal ate 25,634 grams of fish during the week. If, on the first day of the week, he ate 6,987 grams of fish, how many grams of fish did he eat during the remainder of the week?

Use place value understanding to fluently decompose to smaller units multiple times in any place using the standard subtraction algorithm, and apply the algorithm to solve word problems using tape diagrams.
© 2018 Great Minds®. eureka-math.org

EUREKA
MATH

1. In its three months of summer business, the local ice cream stand had a total of $94,326 in sales. The first month's sales were $24,314, and the second month's sales were $30,867.

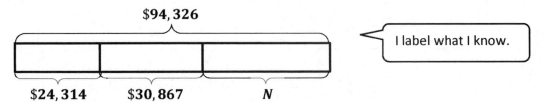

$$\$94,326$$

I label what I know.

$$\$24,314 \qquad \$30,867 \qquad N$$

a. Round each value to the nearest ten thousand to estimate the sales of the third month.

$$\$24,314 \approx \$20,000 \qquad\qquad \$20,000 + \$30,000 = \$50,000$$

$$\$30,867 \approx \$30,000 \qquad\qquad \$90,000 - \$50,000 = \$40,000$$

$$\$94,326 \approx \$90,000 \qquad\qquad \textbf{\textit{The sales of the third month were about }} \$40,000.$$

To estimate the sales of the third month, I subtract the sum from two months from the total amount.

b. Find the exact amount of sales of the third month.

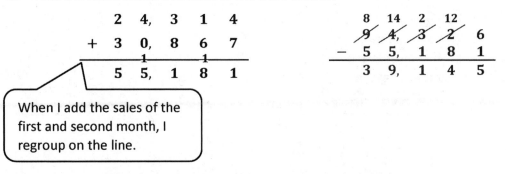

```
      2   4,  3   1   4
  +   3   0,  8   6   7
      ___1_____1____
      5   5,  1   8   1
```

```
      8   14   2   12
      9    4,  3    2   6
  -   5    5,  1    8   1
      _____
      3    9,  1    4   5
```

When I add the sales of the first and second month, I regroup on the line.

The exact amount of sales of the third month was $39,145.

c. Use your answer from part (a) to explain why your answer in part (b) is reasonable.

My answer of $39,145 is reasonable because it is close to my estimate of $40,000. The difference between the actual answer and my estimate is less than $1,000.

EUREKA MATH

Lesson 16: Solve two-step word problems using the standard subtraction algorithm fluently modeled with tape diagrams, and assess the reasonableness of answers using rounding.

© 2018 Great Minds®. eureka-math.org

75

2. In the first month after its release, 55,316 copies of a best-selling book were sold. In the second month after its release, 16,427 fewer copies were sold. How many copies were sold in the first two months? Is your answer reasonable?

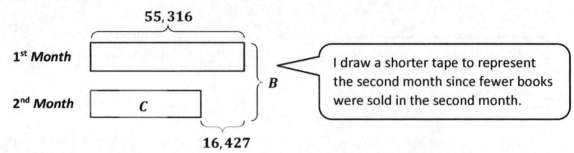

55,316

1st Month

2nd Month   C

B

16,427

> I draw a shorter tape to represent the second month since fewer books were sold in the second month.

Sample Student A Response:

$C = 55,316 - 16,427$

$C = 38,889$

> I subtract to find the actual number of copies sold in the second month.

```
     14 12 10
  4   4  2  0  16
  5   5, 3  1  6
- 1   6, 4  2  7
  3   8, 8  8  9
```

$B = 55,316 + 38,889$

$B = 94,205$

```
  5  5, 3  1  6
+ 3  8, 8  8  9
  1  1  1  1
  9  4, 2  0  5
```

> Then, I add the number of copies of the first and second month together to find the total.

$55,316 \approx 60,000$

$16,427 \approx 20,000$

$60,000 - 20,000 = 40,000$

$60,000 + 40,000 = 100,000$

Sample Student B Response:

$B = 55,316 + 55,316 - 16,427$

$B = 110,632 - 16,427$

$B = 94,205$

```
  5  5, 3  1  6
+ 5  5, 3  1  6
     1        1
1 1  0, 6  3  2
```

> To find the total number of copies I can add two units of 55,316 and then subtract 16,427.

```
        10
  0   0  10    2  12
  1   1  0, 6  3  2
-     1  6, 4  2  7
      9  4, 2  0  5
```

$110,632 \approx 111,000$

$16,427 \approx 16,000$

$111,000 - 16,000 = 95,000$

94,205 copies were sold in the first two months.

> I round to the nearest ten thousand. My answer is reasonable. It is about 6,000 less than my estimate. I would expect this difference because I rounded each number *up* to the nearest ten thousand.

> I round to the nearest thousand. My answer is really close to my estimate! When I round to a smaller place value unit, I often get an estimate closer to the actual answer.

Lesson 16:   Solve two-step word problems using the standard subtraction algorithm fluently modeled with tape diagrams, and assess the reasonableness of answers using rounding.
© 2018 Great Minds®. eureka-math.org

EUREKA MATH

Name _____   Date _____

1. Zachary's final project for a college course took a semester to write and had 95,234 words. Zachary wrote 35,295 words the first month and 19,240 words the second month.

    a. Round each value to the nearest ten thousand to estimate how many words Zachary wrote during the remaining part of the semester.

    b. Find the exact number of words written during the remaining part of the semester.

    c. Use your answer from (a) to explain why your answer in (b) is reasonable.

2. During the first quarter of the year, 351,875 people downloaded an app for their smartphones. During the second quarter of the year, 101,949 fewer people downloaded the app than during the first quarter. How many downloads occurred during the two quarters of the year?

   a. Round each number to the nearest hundred thousand to estimate how many downloads occurred during the first two quarters of the year.

   b. Determine exactly how many downloads occurred during the first two quarters of the year.

   c. Determine if your answer is reasonable. Explain.

Lesson 16: Solve two-step word problems using the standard subtraction algorithm fluently modeled with tape diagrams, and assess the reasonableness of answers using rounding.

© 2018 Great Minds®. eureka-math.org

EUREKA MATH

3. A local store was having a two-week Back to School sale. They started the sale with 36,390 notebooks. During the first week of the sale, 7,424 notebooks were sold. During the second week of the sale, 8,967 notebooks were sold. How many notebooks were left at the end of the two weeks? Is your answer reasonable?

Lesson 16:   Solve two-step word problems using the standard subtraction algorithm fluently modeled with tape diagrams, and assess the reasonableness of answers using rounding.

© 2018 Great Minds®. eureka-math.org

79

Draw a tape diagram to represent each problem. Use numbers to solve, and write your answer as a statement.

1. Saisha has 1,025 stickers. Evan only has 862 stickers. How many more stickers does Saisha have than Evan?

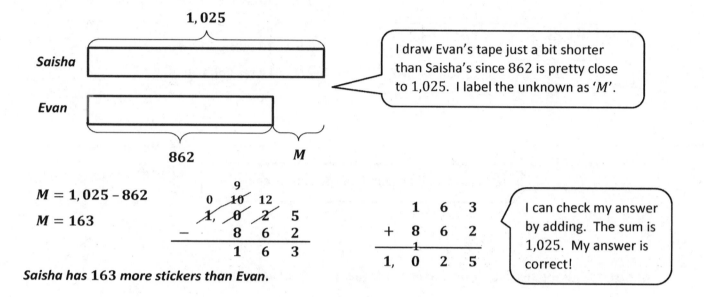

$M = 1,025 - 862$

$M = 163$

Saisha has **163** *more stickers than Evan.*

> I draw Evan's tape just a bit shorter than Saisha's since 862 is pretty close to 1,025. I label the unknown as 'M'.

> I can check my answer by adding. The sum is 1,025. My answer is correct!

2. Milk Truck B contains 3,994 gallons of milk. Together, Milk Truck A and Milk Truck B contain 8,789 gallons of milk. How many more gallons of milk does Milk Truck A contain than Milk Truck B?

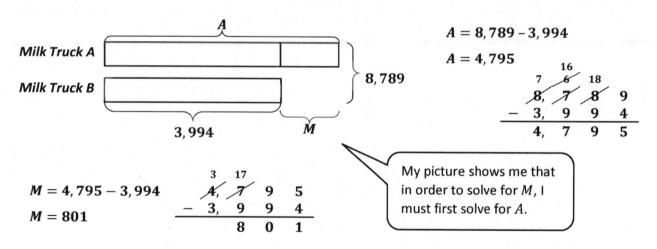

$A = 8,789 - 3,994$

$A = 4,795$

$M = 4,795 - 3,994$

$M = 801$

> My picture shows me that in order to solve for M, I must first solve for A.

Milk Truck A contains 801 more gallons of milk than Milk Truck B.

3. The length of the purple streamer measured 180 inches. After 40 inches were cut from it, the purple streamer was twice as long as the blue streamer. At first, how many inches longer was the purple streamer than the blue streamer?

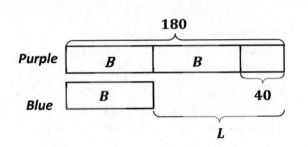

I use unit language to help me solve. The purple streamer is now 140 inches long.

$2B = 18$ tens $- 4$ tens

$2B = 14$ tens or 140

$B = 14$ tens $\div 2$

$B = 7$ tens

$B = 70$

I divide to find the length of the blue streamer.

$L = 180 - 70$

$L = 18$ tens $- 7$ tens

$L = 11$ tens

$L = 110$

At first, the purple streamer was 110 inches longer than the blue streamer.

I subtract the length of the blue streamer from the original length of the purple streamer.

Lesson 17:   Solve *additive compare* word problems modeled with tape diagrams.

EUREKA MATH

Name _____   Date _____

Draw a tape diagram to represent each problem. Use numbers to solve, and write your answer as a statement.

1. Gavin has 1,094 toy building blocks. Avery only has 816 toy building blocks. How many more building blocks does Gavin have?

2. Container B holds 2,391 liters of water. Together, Container A and Container B hold 11,875 liters of water. How many more liters of water does Container A hold than Container B?

Lesson 17:   Solve *additive compare* word problems modeled with tape diagrams.

© 2018 Great Minds®. eureka-math.org

83

3. A piece of yellow yarn was 230 inches long. After 90 inches had been cut from it, the piece of yellow yarn was twice as long as a piece of blue yarn. At first, how much longer was the yellow yarn than the blue yarn?

Lesson 17:   Solve *additive compare* word problems modeled with tape diagrams.

EUREKA
MATH

Draw a tape diagram to represent each problem. Use numbers to solve, and write your answer as a statement.

1. Bridget wrote down three numbers. The first number was $7,401$. The second number was $4,610$ less than the first. The third number was $2,842$ greater than the second. What is the sum of her numbers?

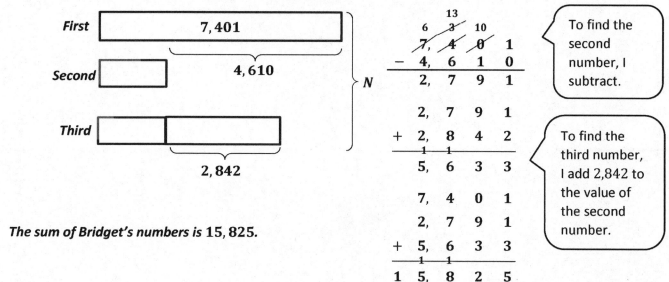

The sum of Bridget's numbers is $15,825$.

2. Mrs. Sample sold a total of 43,210 pounds of mulch. She sold $13,305$ pounds of cherry mulch. She sold $4,617$ more pounds of birch mulch than cherry. The rest of the mulch sold was maple. How many pounds of maple mulch were sold?

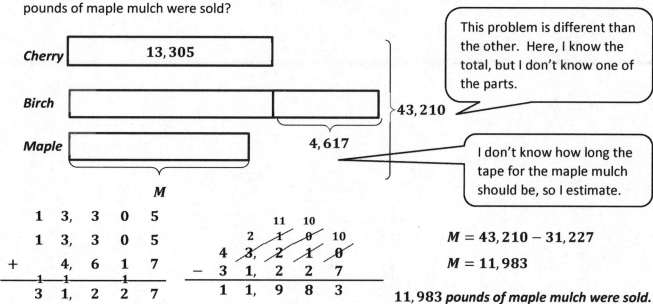

$M = 43,210 - 31,227$

$M = 11,983$

$11,983$ *pounds of maple mulch were sold.*

EUREKA MATH

Lesson 18: Solve multi-step word problems modeled with tape diagrams, and assess the reasonableness of answers using rounding.

85

© 2018 Great Minds®. eureka-math.org

Name _____   Date _____

Draw a tape diagram to represent each problem. Use numbers to solve, and write your answer as a statement.

1. There were 22,869 children, 49,563 men, and 2,872 more women than men at the fair. How many people were at the fair?

2. Number A is 4,676. Number B is 10,043 greater than A. Number C is 2,610 less than B. What is the total value of numbers A, B, and C?

Lesson 18:   Solve multi-step word problems modeled with tape diagrams, and assess the reasonableness of answers using rounding.

© 2018 Great Minds®. eureka-math.org

87

3. A store sold a total of 21,650 balls. It sold 11,795 baseballs. It sold 4,150 fewer basketballs than baseballs. The rest of the balls sold were footballs. How many footballs did the store sell?

Lesson 18:   Solve multi-step word problems modeled with tape diagrams, and assess the reasonableness of answers using rounding.

© 2018 Great Minds®. eureka-math.org

EUREKA
MATH®

1. Using the diagram below, create your own word problem. Solve for the value of the variable, T.

There are 28,596___***people who work for***

Company A. There are 26,325 more ***people***

who work for Company B than Company A.

How many ***people work for the two companies in***
all ?

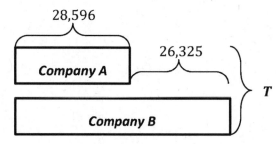

28,596

26,325

Company A

Company B

T

> After analyzing the tape diagram, I create a context for a word problem and
> fill in the blanks. I write "how many in all" because the total, T, is unknown.

Company B = 28,596 + 26,325

```
    2  8,  5  9  6
 +  2  6,  3  2  5
    1      1  1
    5  4,  9  2  1
```

83,517 people work for the two companies in all.

T = Company A + Company B

```
    5  4,  9  2  1
 +  2  8,  5  9  6
    1      1  1
    8  3,  5  1  7
```

2. Use the following tape diagram to create a word problem. Solve for the value of the variable, A.

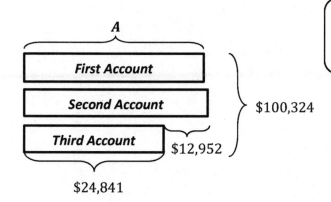

A

First Account

Second Account

Third Account

$100,324

$12,952

$24,841

> I analyze the tape diagram. I find a context,
> and write a word problem based on what is
> known and what is unknown. I label the parts.

Mr. W had 3 bank accounts with a total balance of
$100,324. He had $24,841 in his third account
and $12,952 more in his second account than in his
third account. What was the balance of Mr. W's
first account?

```
    1  2,  9  5  2
 +  2  4,  8  4  1
              1
    3  7,  7  9  3
```

```
    3  7,  7  9  3
 +  2  4,  8  4  1
    1      1  1
    6  2,  6  3  4
```

```
         9   9
        10  10  12  12
    1   0   0,  3   2   4
 -      6   2,  6   3   4
        3   7,  6   9   0
```

Mr. W's first account had a balance of $37,690.

Name _____   Date _____

Using the diagrams below, create your own word problem. Solve for the value of the variable.

1.   At the local botanical gardens, there are _____

Redwoods and _____ Cypress trees.

There are a total of _____ Redwood,

Cypress, and Dogwood trees.

How many _____

_____

_____?

Redwood	Cypress	Dogwood
6,294	3,849	A

12,115

2.   There are 65,302 _____

_____.

There are 37,436 fewer _____

_____.

How many _____

_____?

65,302

37,436

T

EUREKA
MATH®

Lesson 19:   Create and solve multi-step word problems from given tape diagrams
and equations.

© 2018 Great Minds®. eureka-math.org

91

3. Use the following tape diagram to create a word problem. Solve for the value of the variable.

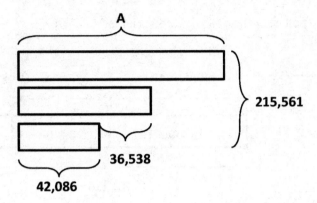

A

215,561

36,538

42,086

4. Draw a tape diagram to model the following equation. Create a word problem. Solve for the value of the variable.

$$27,894 + A + 6,892 = 40,392$$

Lesson 19:   Create and solve multi-step word problems from given tape diagrams and equations.

© 2018 Great Minds®. eureka-math.org

EUREKA
MATH®

Grade 4
Module 2

1. Find the equivalent measures.

   a. 3 km = __**3,000**__ m   b. 4 m = __**400**__ cm

   > I know that 1 kilometer equals 1,000 meters.

   > I know that 1 meter equals 100 centimeters.

2. Find the equivalent measures.

   a. 2 km 345 m = __**2,345**__ m   b. 4 m 23 cm = __**423**__ cm

   c. 12 km 45 m = __**12,045**__ m   d. 24 m 3 cm = __**2,403**__ cm

   > I know that 12 kilometers equals 12,000 meters, so I add 12,000 meters plus 45 meters.

   > I know that 24 meters equals 2,400 centimeters, so I add 2,400 meters plus 3 centimeters.

3. Solve.

   a. 3 m − 42 cm

   Sample Student A Response:

   $$3\text{ m} = 300\text{ cm}$$

   $$
   \begin{array}{r}
   \overset{2}{\cancel{3}}\ \overset{9}{\cancel{0}}\ \overset{10}{\cancel{0}}\ \text{cm}\\
   -\ \ \ \ 4\ \ 2\ \ \text{cm}\\
   \hline
   2\ \ 5\ \ 8\ \ \text{cm}
   \end{array}
   $$

   > Before subtracting, I make like units. 3 meters is equal to 300 centimeters.

   > I'll use the arrow way to add up. I add centimeters and meters that make the next whole.

   Sample Student B Response:

   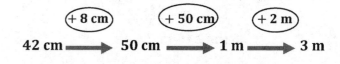

   $$8\text{ cm} + 50\text{ cm} + 2\text{ m} = 2\text{ m }58\text{ cm}$$

   > I add 8 cm to make the next ten, 50 cm. I add 50 cm to make the next meter, and 1 meter is 2 meters away from 3 meters.

   > Now I'll add all the parts circled, finding 2 meters 58 centimeters is the difference of 3 meters and 42 centimeters.

Lesson 1:   Express metric length measurements in terms of a smaller unit;
              model and solve addition and subtraction word problems involving
              metric length.

95

b.   32 m 14 cm − 8 m 63 cm

Sample Student A Response:

```
   2   11      0   11
   ̸3   ̸1      ̸1   ̸1   4
   3   ̸2  m    1   ̸4   cm
 −     8  m        6   3   cm
 _____
       2   3  m    5   1   cm
```

14 cm is not enough to take away 63 cm, so I rename 1 meter as 100 cm to make 114 cm.

Sample Student B Response:

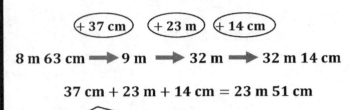

$+ 37$ cm $+ 23$ m $+ 14$ cm

8 m 63 cm ➡ 9 m ➡ 32 m ➡ 32 m 14 cm

37 cm + 23 m + 14 cm = 23 m 51 cm

Using the arrow way, I'll add up from 8 m 63 cm until I reach 32 m 14 cm. It's almost like a number line!

c.   3 km 742 m + 9 km 473 m

Sample Student A Response:

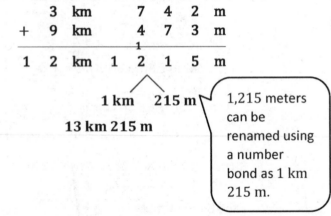

```
     3   km      7   4   2   m
 +   9   km      4   7   3   m
 _____1_____
   1 2   km    1 2   1   5   m
```

1 km   215 m

13 km 215 m

1,215 meters can be renamed using a number bond as 1 km 215 m.

Sample Student B Response:

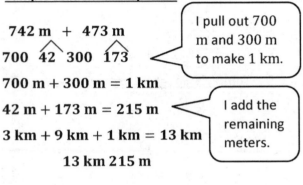

742 m + 473 m

700 42   300 173

700 m + 300 m = 1 km

42 m + 173 m = 215 m

3 km + 9 km + 1 km = 13 km

13 km 215 m

I pull out 700 m and 300 m to make 1 km.

I add the remaining meters.

Use a tape diagram to model each problem. Solve using a simplifying strategy or an algorithm, and write your answer as a statement.

4.   Kya's mom drove 4 km 231 m from work to the grocery store. She drove some more miles from the grocery store to her house. If she drove a total of 8 km, how far was it from her work to her house?

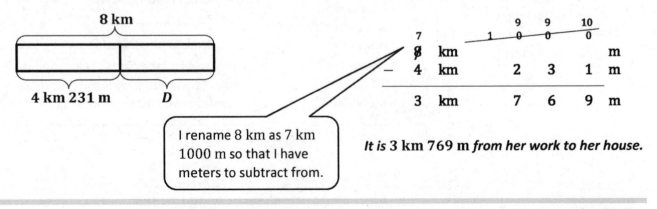

8 km

4 km 231 m   D

```
               9   9   10
       7       ̸1   ̸0   ̸0   ̸0
       ̸8   km                m
   −   4   km       2   3   1   m
   _____
       3   km       7   6   9   m
```

I rename 8 km as 7 km 1000 m so that I have meters to subtract from.

It is 3 km 769 m from her work to her house.

EUREKA MATH®

Name _____   Date _____

1. Find the equivalent measures.

    a. 5 km = _____ m

    b. 13 km = _____ m

    c. _____ km = 17,000 m

    d. 60 km = _____ m

    e. 7 m = _____ cm

    f. 19 m = _____ cm

    g. _____ m = 2,400 cm

    h. 90 m = _____ cm

2. Find the equivalent measures.

    a. 7 km 123 m = _____ m

    b. 22 km 22 m = _____ m

    c. 875 km 4 m = _____ m

    d. 7 m 45 cm = _____ cm

    e. 67 m 7 cm = _____ cm

    f. 204 m 89 cm = _____ cm

3. Solve.

    a. 2 km 303 m – 556 m

    b. 2 m – 54 cm

    c. Express your answer in the smaller unit:
        338 km 853 m + 62 km 71 m

    d. Express your answer in the smaller unit:
        800 m 35 cm – 154 m 49 cm

    e. 701 km – 523 km 445 m

    f. 231 km 811 m + 485 km 829 m

EUREKA MATH

Lesson 1:   Express metric length measurements in terms of a smaller unit;
model and solve addition and subtraction word problems involving
metric length.

© 2018 Great Minds®. eureka-math.org

97

Use a tape diagram to model each problem. Solve using a simplifying strategy or an algorithm, and write your answer as a statement.

4. The length of Celia's garden is 15 m 24 cm. The length of her friend's garden is 2 m 98 cm more than Celia's. What is the length of her friend's garden?

5. Sylvia ran 3 km 290 m in the morning. Then, she ran some more in the evening. If she ran a total of 10 km, how far did Sylvia run in the evening?

6. Jenny's sprinting distance was 356 meters shorter than Tyler's. Tyler sprinted a distance of 1 km 3 m. How many meters did Jenny sprint?

7. The electrician had 7 m 23 cm of electrical wire. He used 551 cm for one wiring project. How many centimeters of wire does he have left?

Lesson 1: Express metric length measurements in terms of a smaller unit; model and solve addition and subtraction word problems involving metric length.

© 2018 Great Minds®. eureka-math.org

1. Complete the conversion table.

Mass	
kg	**g**
3	**3,000**
5	**5,000**
7	7,000

I know that 1 kilogram equals 1,000 grams.

2. Convert the measurements.

a. 4 kg 650 g   = __**4,650**__ g

b. __**51**__ kg __**45**__ g = 51,045 g

In 51,945, there are 51 thousands 945 ones. 1 thousand grams equals 1 kilogram, so 51 thousand grams 945 grams equals 51 kilograms 945 grams.

3. Solve.

a. 7 kg − 860 g

I make like units. 7 kilograms is equal to 7,000 grams.

$$7 \text{ kg} = 7,000 \text{ g}$$

Sample Student A Response:

```
        9
     6  ⅩØ  10
   7,  Ø  Ø  0   g
 −     8  6  0   g
 ─────────────────
   6,  1  4  0   g
```

I subtract grams from grams.

Sample Student B Response:

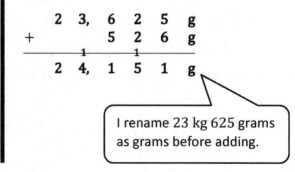

$$860 \text{ g} \longrightarrow 900 \text{ g} \longrightarrow 1,000 \text{ g} \longrightarrow 7,000 \text{ g}$$

$$40 \text{ g} + 100 \text{ g} + 6,000 \text{ g} = 6,140 \text{ g}$$

Just like in Lesson 1, I add up using the arrow way.

b. Express the answer in the smaller unit: 23 kg 625 g + 526 g.

Sample Student A Response:

```
     2  3  kg     6  2  5  g
 +               5  2  6  g
                    1
 ─────────────────────────────
     2  3  kg  1  1  5  1  g
```

I add and then convert the answer to grams.

$$23 \text{ kg} = 23,000 \text{ g}$$

$$23\ 000\ + 1\ 151 \qquad 24\ 151$$

Sample Student B Response:

```
     2  3,  6  2  5  g
 +           5  2  6  g
             1     1
 ────────────────────────
     2  4,  1  5  1  g
```

I rename 23 kg 625 grams as grams before adding.

Lesson 2: Express metric mass measurements in terms of a smaller unit; model and solve addition and subtraction word problems involving metric mass.

© 2018 Great Minds®. eureka-math.org

c.   Express the answer in mixed units: 18 kg 604 g − 3,461 g.

$$3,461\ g = 3\ kg\ 461\ g$$

```
              5  10
    1   8  kg  6̸  0̸  4  g          ┌─────────────────────────┐
  − 3 kg     4   6   1  g          │ I convert grams to kilograms │
  ─────────────────────────        │ before subtracting.          │
    1   5  kg  1   4   3  g        └─────────────────────────┘
```

Use a tape diagram to model each problem. Solve using a simplifying strategy or an algorithm, and write your answer as a statement.

4.   One crate of watermelon weighs 18 kilograms 685 grams. Another crate of watermelon weighs 17 kilograms 435 grams. What is their combined weight?

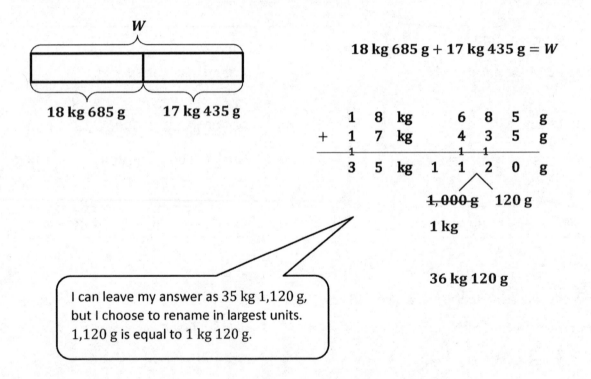

$$18\ kg\ 685\ g + 17\ kg\ 435\ g = W$$

```
    1   8  kg      6   8   5  g
  + 1   7  kg      4   3   5  g
  ─────────────────────────────
    1                 1   1
    3   5  kg    1   1   2   0  g
```

1,000 g   120 g

1 kg

┌──────────────────────────────────┐
│ I can leave my answer as 35 kg 1,120 g, │
│ but I choose to rename in largest units. │
│ 1,120 g is equal to 1 kg 120 g.   │
└──────────────────────────────────┘

36 kg 120 g

The combined weight of the crates of watermelon is 36 kg 120 g.

Lesson 2:   Express metric mass measurements in terms of a smaller unit; model and solve addition and subtraction word problems involving metric mass.
© 2018 Great Minds®. eureka-math.org

EUREKA MATH®

Name _____   Date _____

1. Complete the conversion table.

Mass	
kg	g
1	1,000
6	
	8,000
15	
	24,000
550	

2. Convert the measurements.

a. 2 kg 700 g = _____ g

b. 5 kg 945 g = _____ g

c. 29 kg 58 g = _____ g

d. 31 kg 3 g = _____ g

e. 66,597 g = _____ kg _____ g

f. 270 kg 41 g = _____ g

3. Solve.
   a. 370 g + 80 g

   b. 5 kg – 730 g

   c. Express the answer in the smaller unit:
   27 kg 547 g + 694 g

   d. Express the answer in the smaller unit:
   16 kg + 2,800 g

   e. Express the answer in mixed units:
   4 kg 229 g – 355 g

   f. Express the answer in mixed units:
   70 kg 101 g – 17 kg 862 g

EUREKA MATH

Lesson 2: Express metric mass measurements in terms of a smaller unit; model and solve addition and subtraction word problems involving metric mass.

© 2018 Great Minds®. eureka-math.org

101

Use a tape diagram to model each problem. Solve using a simplifying strategy or an algorithm, and write your answer as a statement.

4. One suitcase weighs 23 kilograms 696 grams. Another suitcase weighs 25 kilograms 528 grams. What is the total weight of the two suitcases?

5. A bag of potatoes and a bag of onions combined weigh 11 kilograms 15 grams. If the bag of potatoes weighs 7 kilograms 300 grams, how much does the bag of onions weigh?

6. The table to the right shows the weight of three dogs. What is the difference in weight between the heaviest and lightest dog?

Dog	Weight
Lassie	21 kg 249 g
Riley	23 kg 128 g
Fido	21,268 g

Lesson 2: Express metric mass measurements in terms of a smaller unit; model and solve addition and subtraction word problems involving metric mass.

EUREKA MATH®

1. Complete the conversion table.

Liquid Capacity	
L	**mL**
6	**6,000**
18	**18,000**
32	32,000

> There are 1,000 milliliters in 1 liter. The rule for converting is the same from Lesson 1 and 2.

2. Convert the measurements.

   a.   26 L 38 mL   =   __26,038__ mL

   b.   427,009 mL   =   __427__ L __9__ mL

> I remember doing these conversions in Lessons 1 and 2, just with different units.

3. Solve.

   a.   Express the answer in the smaller unit:

   32 L 420 mL + 685 mL

```
      3 2,  4  2  0   mL
   +        6  8  5   mL
   ─────1───1──────────
      3 3,  1  0  5   mL
```

> Before adding, I rename 32 L 420 mL as milliliters since the answer is to be in the smaller unit.

   b.   Express the answer in mixed units:

   62 L 608 mL − 35 L 739 mL

```
                     15
   5  11        0  5   9  18
   6   7        1  6   0   8
   6   2  L     6   0   8  mL
 − 3   5  L     7   3   9  mL
   ──────────   ─────────────
   2   6  L     8   6   9  mL
```

> I can subtract mixed units as given, or I can rename the units to the smallest unit, subtract, and then rename as mixed units.

EUREKA MATH

Lesson 3:   Express metric capacity measurements in terms of a smaller unit; model and solve addition and subtraction word problems involving metric capacity.

© 2018 Great Minds®. eureka-math.org

103

Name _____   Date _____

1. Complete the conversion table.

Liquid Capacity	
L	mL
1	1,000
8	
27	
	39,000
68	
	102,000

2. Convert the measurements.

a.   5 L 850 mL = _____ mL

b.   29 L 303 mL = _____ mL

c.   37 L 37 mL = _____ mL

d.   17 L 2 mL = _____ mL

e.   13,674 mL = _____ L _____ mL

f.   275,005 mL = _____ L _____ mL

3. Solve.

a. 545 mL + 48 mL

b. 8 L – 5,740 mL

c. Express the answer in the smaller unit:
    27 L 576 mL + 784 mL

d. Express the answer in the smaller unit:
    27 L + 3,100 mL

e. Express the answer in mixed units:
    9 L 213 mL – 638 mL

f. Express the answer in mixed units:
    41 L 724 mL – 28 L 945 mL

Lesson 3:   Express metric capacity measurements in terms of a smaller unit;
             model and solve addition and subtraction word problems involving
             metric capacity.

© 2018 Great Minds®. eureka-math.org

105

Use a tape diagram to model each problem. Solve using a simplifying strategy or an algorithm, and write your answer as a statement.

4. Sammy's bucket holds 2,530 milliliters of water. Marie's bucket holds 2 liters 30 milliliters of water. Katie's bucket holds 2 liters 350 milliliters of water. Whose bucket holds the least amount of water?

5. At football practice, the water jug was filled with 18 liters 530 milliliters of water. At the end of practice, there were 795 milliliters left. How much water did the team drink?

6. 27,545 milliliters of gas were added to a car's empty gas tank. If the gas tank's capacity is 56 liters 202 milliliters, how much gas is needed to fill the tank?

Lesson 3: Express metric capacity measurements in terms of a smaller unit; model and solve addition and subtraction word problems involving metric capacity.

© 2018 Great Minds®. eureka-math.org

EUREKA MATH

1. Complete the table.

Smaller Unit	Larger Unit	How Many Times as Large as?
ten	thousand	100

> I ask myself, "One thousand is 100 times as large as what unit?" I know 1 thousand is 100 tens (1 × 100 tens). So, my smaller unit is ten.

2. Fill in the unknown unit in word form.

125 is 1 _____*hundred*_____ 25 ones.

> I ask myself, "125 ones is the same as 1 of what larger unit and 25 ones?"

125 cm is 1 _____*meter*_____ 25 cm.

> The units are centimeters. I can make a larger unit. 100 centimeters equals 1 meter. So, 1 meter 25 cm is the same as 125 cm.

3. Write the unknown number.

_____**142,728**_____ is 142 thousands 728 ones.

> I can decompose 142 thousands 728 into smaller units. 142 thousands is the same as 142,000 ones. So, 142 thousands 728 ones is 142,728.

_____**142,728**_____ mL is 142 L 728 mL.

> I know 1 liter equals 1,000 milliliters. So, 142 liters equals 142,000 milliliters, and 142 liters 728 milliliters equals 142,728 milliliters.

4. Fill in each with >, <, or =.

740,259 mL $\bigcirc{>}$ 74 L 249 mL

> 74 L 249 mL is the same as 74,249 mL. 74 ten thousands is greater than 7 ten thousands.

Lesson 4:   Know and relate metric units to place value units in order to express measurements in different units.

© 2018 Great Minds®. eureka-math.org

107

5. Mikal's backpack weighs 4, 289 grams. Mikal weighs 17 kilograms 989 grams more than his backpack. How much do Mikal and his backpack weigh in all?

1 kg = 1, 000 g

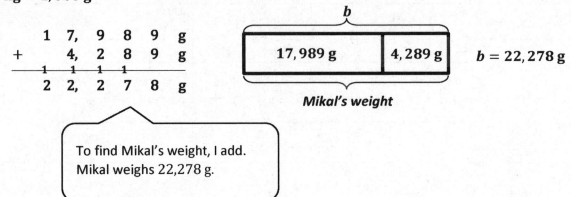

```
  1  7,  9  8  9   g
+     4,  2  8  9   g
   1   1   1   1
  2  2,  2  7  8   g
```

b

| 17, 989 g | 4, 289 g |

b = 22, 278 g

Mikal's weight

To find Mikal's weight, I add. Mikal weighs 22,278 g.

```
  2  2,  2  7  8   g
+     4,  2  8  9   g
          1   1
  2  6,  5  6  7   g
```

c

| 22, 278 g | 4, 289 g |

c = 26, 567 g

Mikal   *backpack*

I add to find the total weight.

Altogether Mikal and his backpack weigh 26, 567 g or 26 kg 567 g.

6. Place the following measurements on the number line:

1 kg 282 g   2,089 g   2 kg 92 g   3,219 g   100 g

Each unit on the number line is 1,000 g. I label each tick mark.

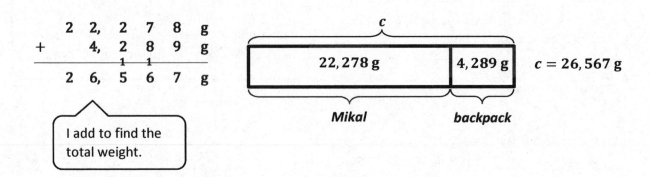

0 g   1,000 g   2,000 g 2,092 g   3,000 g   4,000 g

100 g   1, 282 g   2, 089 g   3, 219 g

I compare 2,092 and 2,089. 9 tens are more than 8 tens. So, 2,092 is more than 2,089.

Lesson 4: Know and relate metric units to place value units in order to express measurements in different units.

Name _____   Date _____

1. Complete the table.

Smaller Unit	Larger Unit	How Many Times as Large as?
centimeter	meter	100
	hundred	100
meter	kilometer	
gram		1,000
one		1,000
milliliter		1,000
one	hundred thousand	

2. Fill in the unknown unit in word form.

   a. 135 is 1 _____35 ones.

   b. 135 cm is 1 _____35 cm.

   c. 1,215 is 1 _____ 215 ones.

   d. 1,215 m is 1 _____215 m.

   e. 12,350 is 12_____350 ones.

   f. 12,350 g is 12 kg 350_____.

3. Write the unknown number.

   a. _____is 125 thousands 312 ones.

   b. _____mL is 125 L 312 mL.

Lesson 4:   Know and relate metric units to place value units in order to express
            measurements in different units.

© 2018 Great Minds®. eureka-math.org

109

2. Box A weighs 30 kilograms 490 grams. Box B weighs 6,790 grams less than Box A. Box C weighs 13 kilograms 757 grams more than Box B. What is the difference, in grams, between the weights of Box C and Box A?

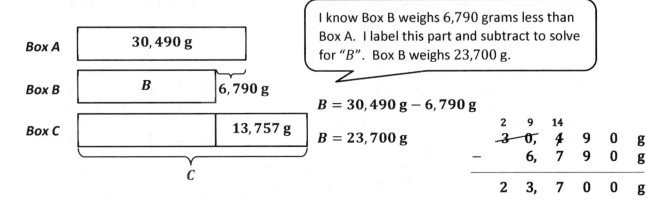

I know Box B weighs 6,790 grams less than Box A. I label this part and subtract to solve for "*B*". Box B weighs 23,700 g.

$$B = 30,490\text{ g} - 6,790\text{ g}$$

$$B = 23,700\text{ g}$$

$$
\begin{array}{r}
{\scriptstyle 2\ \ \ 9\ \ 14}\\
\cancel{3}\ \ \cancel{0},\ \ \cancel{4}\ \ 9\ \ 0\ \ \text{g}\\
-\ \ \ \ \ 6,\ \ 7\ \ 9\ \ 0\ \ \text{g}\\
\hline
2\ \ 3,\ \ 7\ \ 0\ \ 0\ \ \text{g}
\end{array}
$$

I know Box C weighs 13,757 grams more than Box B. If Box B weighs 23,700 grams, I can add to find "C". Box C weighs 37,457 g.

$$
\begin{array}{r}
2\ \ 3,\ \ 7\ \ 0\ \ 0\ \ \text{g}\\
+\ \ 1\ \ 3,\ \ 7\ \ 5\ \ 7\ \ \text{g}\\
{\scriptstyle 1}\\
\hline
3\ \ 7,\ \ 4\ \ 5\ \ 7\ \ \text{g}
\end{array}
$$

I know the weights of Boxes A and C. I can subtract to find the difference, *D*.

$$D = 37,457\text{ g} - 30,490\text{ g}$$

$$D = 6,967\text{ g}$$

$$
\begin{array}{r}
{\scriptstyle 13}\\
{\scriptstyle 6\ \ \ \cancel{3}\ \ 15}\\
3\ \ 7,\ \ \cancel{4}\ \ \cancel{5}\ \ 7\ \ \text{g}\\
-\ \ 3\ \ 0,\ \ 4\ \ 9\ \ 0\ \ \text{g}\\
\hline
6,\ \ 9\ \ 6\ \ 7\ \ \text{g}
\end{array}
$$

The difference between the weights of Box C and Box A is 6,967 g.

Lesson 5:   Use addition and subtraction to solve multi-step word problems involving length, mass, and capacity.

EUREKA MATH

Name _____   Date _____

Model each problem with a tape diagram. Solve and answer with a statement.

1. The capacity of Jose's vase is 2,419 milliliters of water. He poured 1 liter 299 milliliters of water into the empty vase. Then, he added 398 milliliters. How much more water will the vase hold?

2. Eric biked 1 kilometer 125 meters on Monday. On Tuesday, he biked 375 meters less than on Monday. How far did he bike both days?

3. Kristen's package weighs 37 kilograms 95 grams. Anne's package weighs 4,650 grams less than Kristen's. Mary's package weighs 2,905 grams less than Anne's. How much does Mary's package weigh?

Lesson 5:   Use addition and subtraction to solve multi-step word problems involving length, mass, and capacity.

113

4. A Springer Spaniel weighs 20 kilograms 490 grams. A Cocker Spaniel weighs 7,590 grams less than a Springer Spaniel. A Newfoundland weighs 52 kilograms 656 grams more than a Cocker Spaniel. What is the difference, in grams, between the weights of the Newfoundland and the Springer Spaniel?

5. Marsha has three rugs. The first rug is 2 meters 87 centimeters long. The second rug has a length 98 centimeters less than the first. The third rug is 111 centimeters longer than the second rug. What is the difference in centimeters between the length of the first rug and the third rug?

6. One barrel held 60 liters 868 milliliters of sap. A second barrel held 20,089 milliliters more sap than the first. A third barrel held 40 liters 82 milliliters less sap than the second. If the sap from the three barrels was poured into a larger container, how much sap would there be in all?

Lesson 5: Use addition and subtraction to solve multi-step word problems involving length, mass, and capacity.

EUREKA MATH

Grade 4
Module 3

1. Determine the perimeter and area of rectangles A and B.

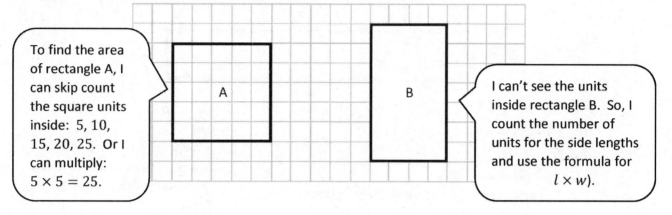

To find the area of rectangle A, I can skip count the square units inside: 5, 10, 15, 20, 25. Or I can multiply: $5 \times 5 = 25$.

I can't see the units inside rectangle B. So, I count the number of units for the side lengths and use the formula for $l \times w$).

a. $A = $ __25 square units__ $A = $ __28 square units__

b. $P = $ __20 units__ $P = $ __22 units__

I can use a formula for perimeter such as $P = 2 \times (l + w)$, $P = l + w + l + w$, or $P = 2l + 2w$.

2. Given the rectangle's area, find the unknown side length.

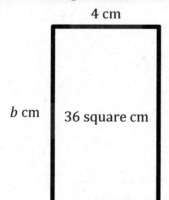

4 cm

b cm   36 square cm

I can think, "4 times what number equals 36?" Or, I can divide to find the unknown side length: $A \div l = w$.

$A = l \times w$
$36 = 4 \times b$
$b = 9$

$b = $ __9__

The unknown side length of the rectangle is 9 centimeters.

EUREKA MATH®

Lesson 1: Investigate and use the formulas for area and perimeter of rectangles.

117

© 2018 Great Minds®. eureka-math.org

3. The perimeter of this rectangle is 250 centimeters. Find the unknown side length of this rectangle.

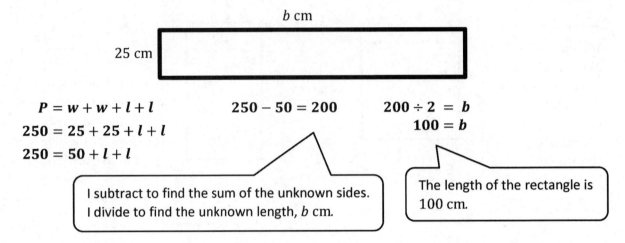

$$P = w + w + l + l$$
$$250 = 25 + 25 + l + l$$
$$250 = 50 + l + l$$

$$250 - 50 = 200$$

$$200 \div 2 = b$$
$$100 = b$$

I subtract to find the sum of the unknown sides.
I divide to find the unknown length, b cm.

The length of the rectangle is 100 cm.

4. The following rectangle has whole number side lengths. Given the area and perimeter, find the length and width.

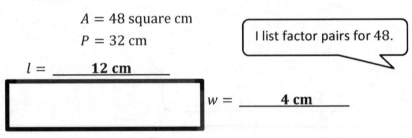

$$A = 48 \text{ square cm}$$
$$P = 32 \text{ cm}$$

I list factor pairs for 48.

$$l = \underline{\quad 12 \text{ cm} \quad}$$

$$w = \underline{\quad 4 \text{ cm} \quad}$$

Dimensions of a
48 square cm *Rectangle*

Width	Length
1 cm	48 cm
2 cm	24 cm
3 cm	16 cm
4 cm	12 cm
6 cm	8 cm

I try the different possible factors as side lengths as I solve for a perimeter of 32 cm using the formula $P = 2L + 2W$.

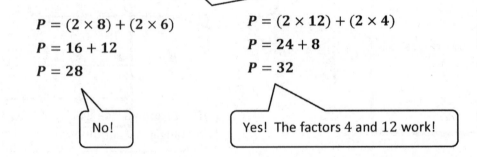

$$P = (2 \times 8) + (2 \times 6)$$
$$P = 16 + 12$$
$$P = 28$$

No!

$$P = (2 \times 12) + (2 \times 4)$$
$$P = 24 + 8$$
$$P = 32$$

Yes! The factors 4 and 12 work!

EUREKA
MATH

Name _____   Date _____

1. Determine the perimeter and area of rectangles A and B.

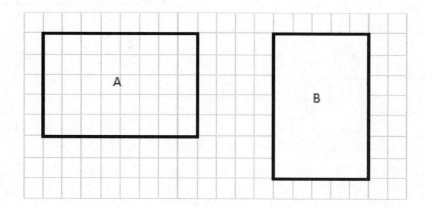

     a.   A = _____   A = _____

     b.   P = _____   P = _____

2. Determine the perimeter and area of each rectangle.

    a.

7 cm

3 cm

P = _____

A = _____

    b.

4 cm

9 cm

P = _____

A = _____

3. Determine the perimeter of each rectangle.

a.

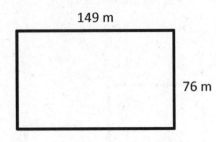

149 m

76 m

P = _____

b.

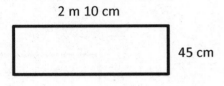

2 m 10 cm

45 cm

P = _____

4. Given the rectangle's area, find the unknown side length.

a.

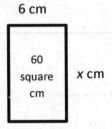

6 cm

60 square cm

x cm

x = _____

b.

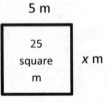

5 m

25 square m

x m

x = _____

Lesson 1:   Investigate and use the formulas for area and perimeter of rectangles.

EUREKA MATH

5. Given the rectangle's perimeter, find the unknown side length.

   a.   P = 180 cm

   40 cm

   x cm

   x = _____

   b.   P = 1,000 m

   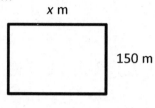

   x m

   150 m

   x = _____

6. Each of the following rectangles has whole number side lengths. Given the area and perimeter, find the length and width.

   a.   A = 32 square cm
        P = 24 cm

   l = _____

   32 square cm

   w = _____

   b.   A = 36 square m
        P = 30 m

   w = _____

   36 square m

   l = _____

1. A rectangular pool is 2 feet wide. It is 4 times as long as it is wide.

    a. Label the diagram with the dimensions of the pool.

    8 ft

    2 ft

    2 ft   2 ft   2 ft   2 ft

    b. Find the perimeter of the pool.

    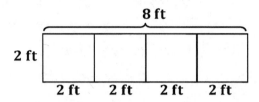

    I choose one of the 3 formulas I learned in Lesson 1 to solve for perimeter.

    $P = 2 \times (l + w)$
    $P = 2 \times (8 + 2)$
    $P = 2 \times 10$
    $P = 20$

    The perimeter of the pool is 20 ft.

2. The area of Brette's bedroom rug is 6 square feet. The longer side measures 3 feet. Her living room rug is twice as long and twice as wide as the bedroom rug.

    a. Draw and label a diagram of Brette's bedroom rug. What is its perimeter?

    3 ft

    b ft   **b ft**

    3 ft

    $A = l \times w$
    $6 = 3 \times w$

    $b = 6 \div 3$ I divide to find the width.
    $b = 2$

    $P = 2l + 2w$

    $P = (2 \times 3) + (2 \times 2)$

    $P = 6 + 4$

    $P = 10$

    The perimeter of Brette's bedroom rug is 10 ft.

EUREKA MATH®   Lesson 2:   Solve multiplicative comparison word problems by applying the area and perimeter formulas.   **123**

© 2018 Great Minds®. eureka-math.org

b.   Draw and label a diagram of Brette's living room rug. What is its perimeter?

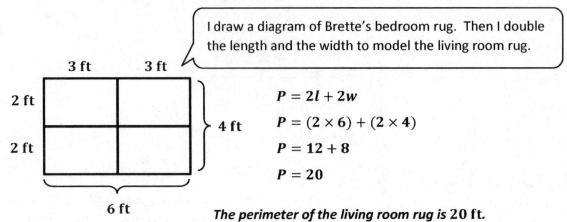

I draw a diagram of Brette's bedroom rug. Then I double the length and the width to model the living room rug.

$P = 2l + 2w$

$P = (2 \times 6) + (2 \times 4)$

$P = 12 + 8$

$P = 20$

The perimeter of the living room rug is 20 ft.

c.   What is the relationship between the two perimeters?

Sample Answer: *The perimeter of the bedroom rug is 10 ft. The perimeter of the living room rug is 20 ft. The living room rug is double the perimeter of the bedroom rug. I know because $2 \times 10 = 20$.*

I explain a pattern I notice. I verify my thinking with an equation.

d.   Find the area of the living room rug using the formula $A = l \times w$.

$A = l \times w$ *The area of the living room rug is 24 square feet.*

$A = 6 \times 4$

$A = 24$

e.   The living room rug has an area that is how many times that of the bedroom rug?

Sample Answer: *The area of the bedroom rug is 6 square feet. The area of the living room rug is 24 square feet. 4 times 6 is 24. The area of the living room rug is 4 times the area of the bedroom rug.*

f.   Compare how the perimeter changed with how the area changed between the two rugs. Explain what you notice using words, pictures, or numbers.

Sample Answer: *The perimeter of the living room rug is 2 times the perimeter of the bedroom rug. But, the area of the living room rug is 4 times the area of the bedroom rug! I notice that when we double each of the side lengths, the perimeter doubles, and the area quadruples.*

     Lesson 2:   Solve multiplicative comparison word problems by applying the area and perimeter formulas.

EUREKA MATH®

Name _____   Date _____

1. A rectangular pool is 7 feet wide. It is 3 times as long as it is wide.

    a. Label the diagram with the dimensions of the pool.

    b. Find the perimeter of the pool.

2. A poster is 3 inches long. It is 4 times as wide as it is long.

    a. Draw a diagram of the poster, and label its dimensions.

    b. Find the perimeter and area of the poster.

Lesson 2:   Solve multiplicative comparison word problems by applying the area
and perimeter formulas.

125

© 2018 Great Minds®. eureka-math.org

3. The area of a rectangle is 36 square centimeters, and its length is 9 centimeters.

   a. What is the width of the rectangle?

   b. Elsa wants to draw a second rectangle that is the same length but is 3 times as wide. Draw and label Elsa's second rectangle.

   c. What is the perimeter of Elsa's second rectangle?

Lesson 2:   Solve multiplicative comparison word problems by applying the area and perimeter formulas.

© 2018 Great Minds®. eureka-math.org

4. The area of Nathan's bedroom rug is 15 square feet. The longer side measures 5 feet. His living room rug is twice as long and twice as wide as the bedroom rug.

    a. Draw and label a diagram of Nathan's bedroom rug. What is its perimeter?

    b. Draw and label a diagram of Nathan's living room rug. What is its perimeter?

    c. What is the relationship between the two perimeters?

    d. Find the area of the living room rug using the formula $A = l \times w$.

Lesson 2:   Solve multiplicative comparison word problems by applying the area and perimeter formulas.

e. The living room rug has an area that is how many times that of the bedroom rug?

f. Compare how the perimeter changed with how the area changed between the two rugs. Explain what you notice using words, pictures, or numbers.

Lesson 2: Solve multiplicative comparison word problems by applying the area and perimeter formulas.

© 2018 Great Minds®. eureka-math.org

EUREKA
MATH

Solve the following problems. Use pictures, numbers, or words to show your work.

A calendar is 2 times as long and 3 times as wide as a business card. The business card is 2 inches long and 1 inch wide. What is the perimeter of the calendar?

1.

2 in
1 in [   ]

2 in   **2 in**

1 in
1 in
1 in

} **3 in**

4 in

$P = 2 \times (l + w)$

$P = 2 \times (4 \text{ in} + 3 \text{ in})$

$P = 2 \times 7 \text{ in}$

$P = 14 \text{ in}$

*The perimeter of the calendar is **14** inches.*

> I draw a diagram with a width 3 times that of the card (3 in).
> I label the length to equal twice the width of the card (4 in).

Rectangle A has an area of 64 square centimeters. Rectangle A is 8 times as many square centimeters as rectangle B. If rectangle B is 4 centimeters wide, what is the length of rectangle B?

> There are so many ways to solve!

2.

64 square

cm

Rectangle A

1 unit = B square cm

8 units = 64 square cm

$64 \div 8 = B$

$B = 8$

> The area of rectangle B is 8 square centimeters.

l

8

square

cm

4 cm

Rectangle B

$A = w \times l$

$8 = 4 \times l$

$l = 8 \div 4$

$l = 2$

*The length of rectangle B is **2** cm.*

EUREKA
MATH®

Lesson 3: Demonstrate understanding of area and perimeter formulas by solving multi-step real-world problems.

129

© 2018 Great Minds®. eureka-math.org

Name _____   Date _____

Solve the following problems. Use pictures, numbers, or words to show your work.

1. Katie cut out a rectangular piece of wrapping paper that was 2 times as long and 3 times as wide as the box that she was wrapping. The box was 5 inches long and 4 inches wide. What is the perimeter of the wrapping paper that Katie cut?

2. Alexis has a rectangular piece of red paper that is 4 centimeters wide. Its length is twice its width. She glues a rectangular piece of blue paper on top of the red piece measuring 3 centimeters by 7 centimeters. How many square centimeters of red paper will be visible on top?

Lesson 3:   Demonstrate understanding of area and perimeter formulas by solving
multi-step real-world problems.

131

3. Brinn's rectangular kitchen has an area of 81 square feet. The kitchen is 9 times as many square feet as Brinn's pantry. If the rectangular pantry is 3 feet wide, what is the length of the pantry?

4. The length of Marshall's rectangular poster is 2 times its width. If the perimeter is 24 inches, what is the area of the poster?

Demonstrate understanding of area and perimeter formulas by solving
multi-step real-world problems.

EUREKA
MATH

1. Fill in the blanks in the following equations.

   a. __**100**__ × 7 = 700

   b. 4 × __**1,000**__ = 4,000

   c. __**50**__ = 10 × 5

   > I ask myself, "How many sevens are equal to 700?"

   > I use unit form to solve. If I name the units, multiplying large numbers is easy! I know 4 ÷ 4 = 1, so 4 thousands ÷ 4 is 1 thousand.

Draw place value disks and arrows to represent each product.

2. 15 × 100 = __**1,500**__

   15 × 10 × 10 = __**1,500**__

   (1 ten 5 ones) × 100 = __**1 thousand 5 hundreds**__

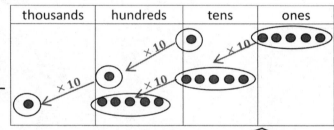

thousands	hundreds	tens	ones

   > Fifteen is 1 ten 5 ones. I draw an arrow to show times 10 for the 1 ten and also for the 5 ones. I multiply by 10 again and I have 1 thousand 5 hundreds.

   > If I shift a digit one place to the left on the chart, that digit becomes 10 times as much as its value to the right.

Decompose each multiple of 10, 100, or 1,000 before multiplying.

3. 2 × 300 = 2 × __**3**__ × __**100**__

   = __**6**__ × __**100**__

   = __**600**__

4. 6 × 7,000 = __**6**__ × __**7**__ × __**1,000**__

   = __**42**__ × __**1,000**__

   = __**42,000**__

   > I can decompose 300 to make an easy fact to solve! I know 2 × 3 hundreds = 6 hundreds.

EUREKA MATH

Lesson 4: Interpret and represent patterns when multiplying by 10, 100, and 1,000 in arrays and numerically.

© 2018 Great Minds®. eureka-math.org

133

Name _____   Date _____

Example:

$5 \times 10 =$ ___50___

5 ones $\times 10 =$ __5__ __tens__

thousands	hundreds	tens	ones

Draw place value disks and arrows as shown to represent each product.

1. $7 \times 100 =$ _____

    $7 \times 10 \times 10 =$ _____

    7 ones $\times 100 =$ _____

thousands	hundreds	tens	ones

2. $7 \times 1{,}000 =$ _____

    $7 \times 10 \times 10 \times 10 =$ _____

    7 ones $\times 1{,}000 =$ ____

    _____

thousands	hundreds	tens	ones

3. Fill in the blanks in the following equations.

    a. $8 \times 10 =$ _____

    b. _____ $\times 8 = 800$

    c. $8{,}000 =$ _____ $\times 1{,}000$

    d. $10 \times 3 =$ _____

    e. $3 \times$ _____ $= 3{,}000$

    f. _____ $\times 3 = 300$

    g. $1{,}000 \times 4 =$ _____

    h. _____ $= 10 \times 4$

    i. $400 =$ _____ $\times 100$

Lesson 4:   Interpret and represent patterns when multiplying by 10, 100, and
            1,000 in arrays and numerically.

© 2018 Great Minds®. eureka-math.org

135

Draw place value disks and arrows to represent each product.

4. 15 × 10 = _____

    (1 ten 5 ones) × 10 = _____

thousands	hundreds	tens	ones

5. 17 × 100 = _____

    17 × 10 × 10 = _____

    (1 ten 7 ones) × 100 = _____

thousands	hundreds	tens	ones

6. 36 × 1,000 = _____

    36 × 10 × 10 × 10 = _____

    (3 tens 6 ones) × 1,000 = _____

ten thousands	thousands	hundreds	tens	ones

Decompose each multiple of 10, 100, or 1000 before multiplying.

7. 2 × 80 = 2 × 8 × _____

    = 16 × _____

    = _____

8. 2 × 400 = 2 × _____ × _____

    = _____ × _____

    = _____

9. 5 × 5,000 = _____ × _____ × _____

    = _____ × _____

    = _____

10. 7 × 6,000 = _____ × _____ × _____

    = _____ × _____

    = _____

Lesson 4: Interpret and represent patterns when multiplying by 10, 100, and 1,000 in arrays and numerically.

EUREKA MATH

1. $2 \times 4,000 =$ __8,000__

    __2__ times __4 thousands__ is __8 thousands__.

	thousands	hundreds	tens	ones
	●●●●			
	●●●●			

> I draw 2 groups of 4 thousands and circle each group. I see a pattern! 2 groups of 4 units is 8 units.

$$4,\ 0\ 0\ 0$$
$$\times \qquad\qquad 2$$
$$\overline{8,\ 0\ 0\ 0}$$

2×4 thousands $= 8$ thousands

> Writing the equation in unit form helps me when one of the factors is a multiple of 10.

2. Find the product.

a. $4 \times 70 = 280$	b. $4 \times 60 = 240$	c. $4 \times 500 = 2,000$	d. $6,000 \times 5 = 30,000$
4×7 tens $= 28$ tens	4×6 tens $= 24$ tens	4×5 hundreds $= 20$ hundreds	6 thousands $\times 5$ $= 30$ thousands

3. At the school cafeteria, each student who orders lunch gets 7 chicken nuggets. The cafeteria staff prepares enough for 400 kids. How many chicken nuggets does the cafeteria staff prepare altogether?

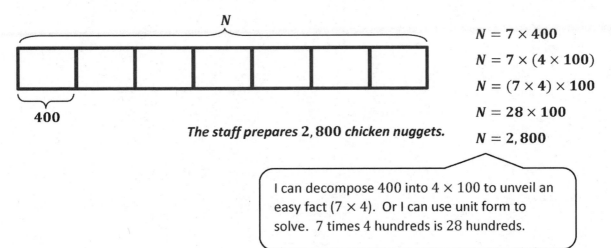

$N = 7 \times 400$

$N = 7 \times (4 \times 100)$

$N = (7 \times 4) \times 100$

$N = 28 \times 100$

$N = 2,800$

The staff prepares 2,800 chicken nuggets.

> I can decompose 400 into 4×100 to unveil an easy fact (7×4). Or I can use unit form to solve. 7 times 4 hundreds is 28 hundreds.

EUREKA MATH

Lesson 5:   Multiply multiples of 10, 100, and 1,000 by single digits, recognizing patterns.

137

© 2018 Great Minds®. eureka-math.org

Name _____   Date _____

Draw place value disks to represent the value of the following expressions.

1.   5 × 2 = _____

5 times _____ ones is _____ ones.

thousands	hundreds	tens	ones

$$\begin{array}{r} 2 \\ \times\ 5 \\ \hline \end{array}$$

2.   5 × 20 = _____

5 times _____ tens is _____.

thousands	hundreds	tens	ones

$$\begin{array}{r} 20 \\ \times\ \ 5 \\ \hline \end{array}$$

3.   5 × 200 = _____

5 times _____ is _____.

thousands	hundreds	tens	ones

$$\begin{array}{r} 200 \\ \times\ \ \ 5 \\ \hline \end{array}$$

4.   5 × 2,000 = _____

_____ times _____ is _____.

thousands	hundreds	tens	ones

$$\begin{array}{r} 2000 \\ \times\ \ \ \ 5 \\ \hline \end{array}$$

Lesson 5:   Multiply multiples of 10, 100, and 1,000 by single digits, recognizing
            patterns.

© 2018 Great Minds®. eureka-math.org

5. Find the product.

a. 20 × 9	b. 6 × 70	c. 7 × 700	d. 3 × 900
e. 9 × 90	f. 40 × 7	g. 600 × 6	h. 8 × 6,000
i. 5 × 70	j. 5 × 80	k. 5 × 200	l. 6,000 × 5

6. At the school cafeteria, each student who orders lunch gets 6 chicken nuggets. The cafeteria staff prepares enough for 300 kids. How many chicken nuggets does the cafeteria staff prepare altogether?

Lesson 5:   Multiply multiples of 10, 100, and 1,000 by single digits, recognizing patterns.

EUREKA
MATH

7. Jaelynn has 30 times as many stickers as her brother. Her brother has 8 stickers. How many stickers does Jaelynn have?

8. The flower shop has 40 times as many flowers in one cooler as Julia has in her bouquet. The cooler has 120 flowers. How many flowers are in Julia's bouquet?

Lesson 5:   Multiply multiples of 10, 100, and 1,000 by single digits, recognizing patterns.

141

© 2018 Great Minds®. eureka-math.org

disks in the place value chart.

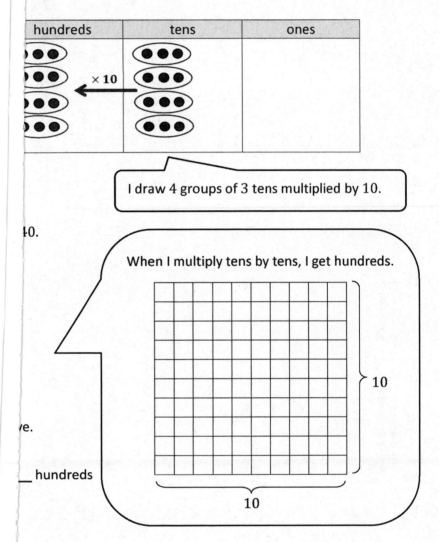

hundreds	tens	ones

I draw 4 groups of 3 tens multiplied by 10.

40.

When I multiply tens by tens, I get hundreds.

10

10

ve.

___ hundreds

are 70 cartons in a crate, how many eggs are in one crate?

7 tens × 7 tens = 49 hundreds

70 × 70 = 4,900

There are 4,900 eggs in one crate.

two-digit multiples of 10 by two-digit multiples of 10 with the
del.

th.org

Name _____   Date _____

Represent the following problem by drawing disks in the place value chart.

1. To solve 30 × 60, think

   (3 tens × 6) × 10 = _____

   30 × (6 × 10) = _____

   30 × 60 = _____

hundreds	tens	ones

2. Draw an area model to represent 30 × 60.

   3 tens × 6 tens = _____ _____

3. Draw an area model to represent 20 × 20.

   2 tens × 2 tens = _____ _____

   20 × 20 = _____

Lesson 6: Multiply two-digit multiples of 10 by two-digit multiples of 10 with the area model.

© 2018 Great Minds®. eureka-math.org

145

4. Draw an area model to represent 40 × 60.

4 tens × 6 tens = _____ _____

40 × 60 = _____

Rewrite each equation in unit form and solve.

5. 50 × 20 = _____

5 tens × 2 tens = _____ hundreds

6. 30 × 50 = _____

3 tens × 5 _____ = _____ hundreds

7. 60 × 20 = _____

_____ tens × _____ tens = 12 _____

8. 40 × 70 = _____

____ _____ × ____ _____ = _____ hundreds

Multiply two-digit multiples of 10 by two-digit multiples of 10 with the area model.

EUREKA
MATH

9. There are 60 seconds in a minute and 60 minutes in an hour. How many seconds are in one hour?

10. To print a comic book, 50 pieces of paper are needed. How many pieces of paper are needed to print 40 comic books?

Lesson 6: Multiply two-digit multiples of 10 by two-digit multiples of 10 with the area model.

© 2018 Great Minds®. eureka-math.org

147

1. Represent the following expression with disks, regrouping as necessary. To the right, record the partial products vertically.

4×35

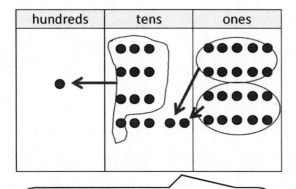

		3	5
×			4
		2	0	→ 4×5 *ones*
+	1	2	0	→ 4×3 *tens*
	1	4	0	

I draw 4 groups of 3 tens 5 ones.

4 times 5 ones equals 20 ones.

I compose 20 ones to make 2 tens.

4 times 3 tens equals 12 tens.

I compose 10 tens to make 1 hundred.

After multiplying the ones, I record the product. I multiply the tens and record the product. I add these two partial products. My sum is the product of 35×4.

2. Jillian says she found a shortcut for doing multiplication problems. When she multiplies 3×45, she says, "3×5 is 15 ones, or 1 ten and 5 ones. Then, there's just 4 tens left in 45, so add it up, and you get 5 tens and 5 ones." Do you think Jillian's shortcut works? Explain your thinking in words, and justify your response using a model or partial products.

Sample answer:

Jillian multiplied the ones. She found the first partial product. But she didn't multiply the tens. She forgot to multiply 4 tens by 3. So, Jillian didn't get the right second partial product. So, her final product isn't correct. The product of 3×45 is 135.

		4	5
×			3
		1	5	→ 3×5 *ones*
+	1	2	0	→ 3×4 *tens*
	1	3	5	

© 2018 Great Minds®. eureka-math.org

Name _____ Date _____

1. Represent the following expressions with disks, regrouping as necessary, writing a matching expression, and recording the partial products vertically.

   a. 3 × 24

tens	ones

   b. 3 × 42

hundreds	tens	ones

   c. 4 × 34

hundreds	tens	ones

EUREKA MATH®

Lesson 7:   Use place value disks to represent two-digit by one-digit multiplication.

151

© 2018 Great Minds®. eureka-math.org

2. Represent the following expressions with disks, regrouping as necessary. To the right, record the partial products vertically.

    a. 4 × 27

hundreds	tens	ones

    b. 5 × 42

hundreds	tens	ones

3. Cindy says she found a shortcut for doing multiplication problems. When she multiplies 3 × 24, she says, "3 × 4 is 12 ones, or 1 ten and 2 ones. Then, there's just 2 tens left in 24, so add it up, and you get 3 tens and 2 ones." Do you think Cindy's shortcut works? Explain your thinking in words, and justify your response using a model or partial products.

EUREKA
MATH®

Represent the following with disks, using either method shown in class, regrouping as necessary. Below the place value chart, record the partial product vertically.

1. 5×731

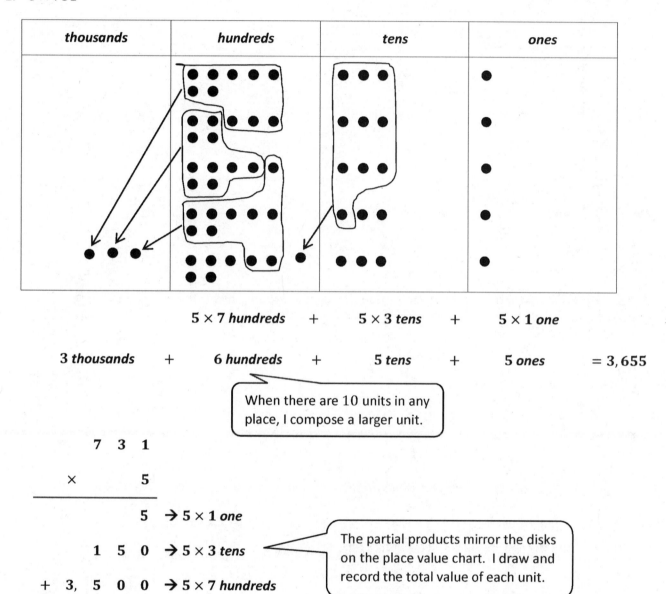

thousands	hundreds	tens	ones

5×7 **hundreds** + 5×3 **tens** + 5×1 **one**

3 **thousands** + 6 **hundreds** + 5 **tens** + 5 **ones** $= 3,655$

> When there are 10 units in any place, I compose a larger unit.

```
      7  3  1
   ×        5
   -----------
            5   → 5 × 1 one

      1  5  0   → 5 × 3 tens

 +  3, 5  0  0  → 5 × 7 hundreds
   -----------
    3, 6  5  5
```

> The partial products mirror the disks on the place value chart. I draw and record the total value of each unit.

2. Janice rides her bike around the block. The block is rectangular with a width of 172 m and a length of 230 m.

   a. Determine how many meters Janice rides if she goes around the block one time.

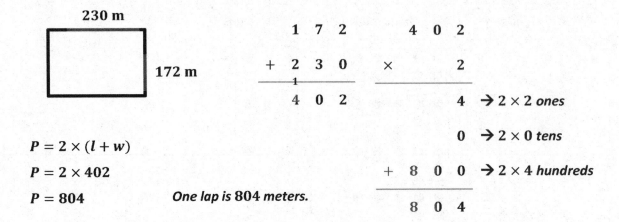

230 m

172 m

$$
\begin{array}{r}
1\ 7\ 2 \\
+\ 2\ 3\ 0 \\
\scriptstyle 1 \\
\hline
4\ 0\ 2
\end{array}
$$

$$P = 2 \times (l + w)$$

$$P = 2 \times 402$$

$$P = 804$$ **One lap is 804 meters.**

$$
\begin{array}{r}
4\ 0\ 2 \\
\times \qquad 2 \\
\hline
4 \quad \rightarrow 2 \times 2\ \textit{ones} \\[4pt]
0 \quad \rightarrow 2 \times 0\ \textit{tens} \\[4pt]
+\ 8\ 0\ 0 \quad \rightarrow 2 \times 4\ \textit{hundreds} \\
\hline
8\ 0\ 4
\end{array}
$$

   b. Determine how many meters Janice rides if she goes around the block three times.

$$
\begin{array}{r}
8\ 0\ 4 \\
\times \qquad 3 \\
\hline
1\ 2 \quad \rightarrow 3 \times 4\ \textit{ones} \\[4pt]
0 \quad \rightarrow 3 \times 0\ \textit{tens} \\[4pt]
+\ 2,\ 4\ 0\ 0 \quad \rightarrow 3 \times 8\ \textit{hundreds} \\
\hline
2,\ 4\ 1\ 2
\end{array}
$$

Janice rides 2,412 meters.

Lesson 8:   Extend the use of place value disks to represent three- and four-digit by one-digit multiplication.

© 2018 Great Minds®. eureka-math.org

EUREKA MATH

Name _____   Date _____

1. Represent the following expressions with disks, regrouping as necessary, writing a matching expression, and recording the partial products vertically as shown below.

   a. 2×424

hundreds	tens	ones
● ● ● ●	● ●	● ● ● ●

   $$\begin{array}{r} 4 \quad 2 \quad 4 \\ \times \quad\quad\quad 2 \\ \hline \end{array}$$

   → $2 \times$ ___ ones

   → $2 \times$ ___ _____

   + _____

   → ___ $\times$ ___ _____

   $2 \times$ ___ _____ + $2 \times$ ___ _____ + $2 \times$ ___ ones

   b. 3×424

hundreds	tens	ones

   c. $4 \times 1,424$

2. Represent the following expressions with disks, using either method shown in class, regrouping as necessary. To the right, record the partial products vertically.

   a. 2 × 617

   b. 5 × 642

   c. 3 × 3,034

Lesson 8: Extend the use of place value disks to represent three- and four-digit by one-digit multiplication.

EUREKA MATH

3. Every day, Penelope jogs three laps around the playground to keep in shape. The playground is rectangular with a width of 163 m and a length of 320 m.

   a. Find the total amount of meters in one lap.

   b. Determine how many meters Penelope jogs in three laps.

Lesson 8:   Extend the use of place value disks to represent three- and four-digit by one-digit multiplication.

157

© 2018 Great Minds®. eureka-math.org

1. Solve using each method.

> No matter which method I choose, I get the same product.

Partial Products	Standard Algorithm
$\begin{array}{r} 2\ \ 1\ \ 5 \\ \times \qquad 4 \\ \hline \ \ 2\ \ 0 \\ \ \ 4\ \ 0 \\ +\ 8\ \ 0\ \ 0 \\ \hline \ 8\ \ 6\ \ 0 \end{array}$	$\begin{array}{r} 2\ \ 1\ \ 5 \\ \times \qquad 4 \\ \hline 8\ \ 6\ \ 0 \end{array}$

> I envision my work with disks on the place value chart when I use the partial products method. I record each partial product on a separate line.

> When using the standard algorithm, I record the product all on one line.

> 4 times 5 ones equals 20 ones or 2 tens 0 ones. I record 2 tens on the line in the tens place and 0 ones in the ones place.

2. Solve using the standard algorithm.

a.
$\begin{array}{r} 2\ \ 0\ \ 5 \\ \times \qquad 9 \\ \hline 1,\ 8\ \ 4\ \ 5 \end{array}$

b.
$\begin{array}{r} 4\ \ 9\ \ 1 \\ \times \qquad 7 \\ \hline 3,\ 4\ \ 3\ \ 7 \end{array}$

> When using the standard algorithm, I multiply the ones first.

> 7 times 4 hundreds is 28 hundreds. I add 6 hundreds and record 34 hundreds. I cross out the 6 hundreds after I add them.

3. One airline ticket costs $249. How much will 4 tickets cost?

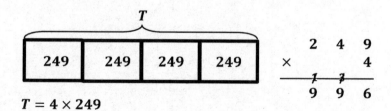

$\begin{array}{r} 2\ \ 4\ \ 9 \\ \times \qquad 4 \\ \hline 9\ \ 9\ \ 6 \end{array}$

> I record 36 ones as 3 tens 6 ones. I write the 3 first and then the 6. It's easy to see 36 since the 3 is written on the line.

$T = 4 \times 249$

$T = 996$

Four tickets will cost $996.

EUREKA MATH   Lesson 9:   Multiply three- and four-digit numbers by one-digit numbers applying the standard algorithm.   159

© 2018 Great Minds®. eureka-math.org

Name _____ Date _____

1. Solve using each method.

Partial Products	Standard Algorithm
a. 4 6 × 2	4 6 × 2

Partial Products	Standard Algorithm
b. 3 1 5 × 4	3 1 5 × 4

2. Solve using the standard algorithm.

a. 2 3 2 × 4	b. 1 4 2 × 6	c. 3 1 4 × 7
d. 4 4 0 × 3	e. 5 0 7 × 8	f. 3 8 4 × 9

EUREKA MATH

Lesson 9: Multiply three- and four-digit numbers by one-digit numbers applying the standard algorithm.

161

© 2018 Great Minds®. eureka-math.org

3. What is the product of 8 and 54?

4. Isabel earned 350 points while she was playing Blasting Robot. Isabel's mom earned 3 times as many points as Isabel. How many points did Isabel's mom earn?

5. To get enough money to go on a field trip, every student in a club has to raise $53 by selling chocolate bars. There are 9 students in the club. How much money does the club need to raise to go on the field trip?

Lesson 9: Multiply three- and four-digit numbers by one-digit numbers applying the standard algorithm.

6. Mr. Meyers wants to order 4 tablets for his classroom. Each tablet costs $329. How much will all four tablets cost?

7. Amaya read 64 pages last week. Amaya's older brother, Rogelio, read twice as many pages in the same amount of time. Their big sister, Elianna, is in high school and read 4 times as many pages as Rogelio did. How many pages did Elianna read last week?

Lesson 9:   Multiply three- and four-digit numbers by one-digit numbers applying the standard algorithm.

163

1. Solve using the standard algorithm.

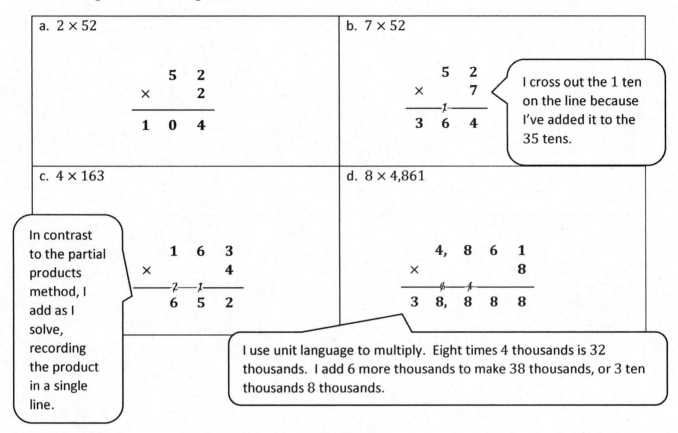

a. 2 × 52

$$
\begin{array}{r}
5\ 2 \\
\times\ \ \ 2 \\
\hline
1\ 0\ 4
\end{array}
$$

b. 7 × 52

$$
\begin{array}{r}
5\ 2 \\
\times\ \ \ 7 \\
\hline
3\ 6\ 4
\end{array}
$$

I cross out the 1 ten on the line because I've added it to the 35 tens.

c. 4 × 163

In contrast to the partial products method, I add as I solve, recording the product in a single line.

$$
\begin{array}{r}
1\ 6\ 3 \\
\times\ \ \ \ \ 4 \\
\hline
6\ 5\ 2
\end{array}
$$

d. 8 × 4,861

$$
\begin{array}{r}
4,\ 8\ 6\ 1 \\
\times\ \ \ \ \ \ \ 8 \\
\hline
3\ 8,\ 8\ 8\ 8
\end{array}
$$

I use unit language to multiply. Eight times 4 thousands is 32 thousands. I add 6 more thousands to make 38 thousands, or 3 ten thousands 8 thousands.

2. Mimi ran 2 miles. Raj ran 3 times as far. There are 5,280 feet in a mile. How many feet did Raj run?

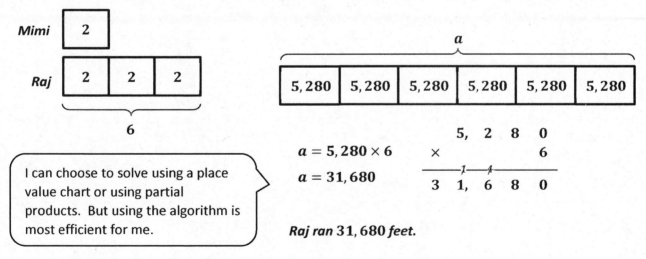

Mimi | 2

Raj | 2 | 2 | 2

6

I can choose to solve using a place value chart or using partial products. But using the algorithm is most efficient for me.

a

| 5,280 | 5,280 | 5,280 | 5,280 | 5,280 | 5,280 |

$a = 5,280 \times 6$

$a = 31,680$

$$
\begin{array}{r}
5,\ 2\ 8\ 0 \\
\times\ \ \ \ \ \ \ 6 \\
\hline
3\ 1,\ 6\ 8\ 0
\end{array}
$$

Raj ran 31,680 feet.

Lesson 10: Multiply three- and four-digit numbers by one-digit numbers applying the standard algorithm.

165

Name _____   Date _____

1. Solve using the standard algorithm.

a. 3 × 41	b. 9 × 41
c. 7 × 143	d. 7 × 286
e. 4 × 2,048	f. 4 × 4,096
g. 8 × 4,096	h. 4 × 8,192

Lesson 10:   Multiply three- and four-digit numbers by one-digit numbers
applying the standard algorithm.

167

© 2018 Great Minds®. eureka-math.org

2. Robert's family brings six gallons of water for the players on the football team. If one gallon of water contains 128 fluid ounces, how many fluid ounces are in six gallons?

3. It takes 687 Earth days for the planet Mars to revolve around the sun once. How many Earth days does it take Mars to revolve around the sun four times?

4. Tammy buys a 4-gigabyte memory card for her camera. Dijonea buys a memory card with twice as much storage as Tammy's. One gigabyte is 1,024 megabytes. How many megabytes of storage does Dijonea have on her memory card?

Lesson 10:   Multiply three- and four-digit numbers by one-digit numbers applying the standard algorithm.

© 2018 Great Minds®. eureka-math.org

1. Solve the following expression using the standard algorithm, the partial products method, and the area model.

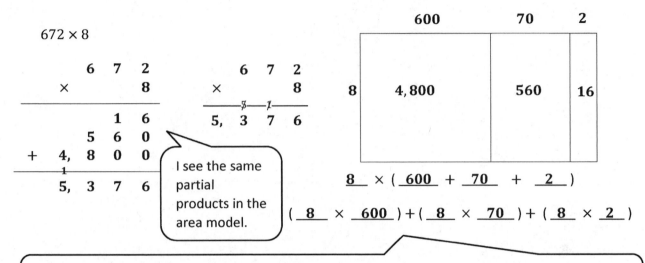

672×8

I see the same partial products in the area model.

$8 \times (\ \underline{600}\ +\ \underline{70}\ +\ \underline{2}\ )$

$(\ \underline{8}\ \times\ \underline{600}\ )+(\ \underline{8}\ \times\ \underline{70}\ )+(\ \underline{8}\ \times\ \underline{2}\ )$

I multiply unit by unit when solving using partial products, the algorithm, or the area model. All along I have been using the distributive property! Now I can write it out as an expression to match.

2. Solve using the standard algorithm, the area model, the distributive property, or the partial products method.

Each year, Mr. Hill gives $5,725 to charity, and Mrs. Hill gives $752. After 5 years, how much has the couple given to charity?

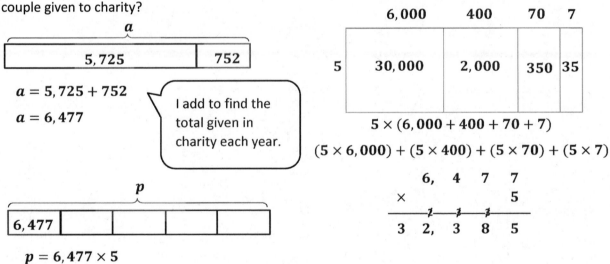

$a = 5,725 + 752$

$a = 6,477$

I add to find the total given in charity each year.

$5 \times (6,000 + 400 + 70 + 7)$

$(5 \times 6,000) + (5 \times 400) + (5 \times 70) + (5 \times 7)$

$p = 6,477 \times 5$

$p = 32,385$

After 5 years, Mr. and Mrs. Hill have given $32,385 to charity.

EUREKA MATH

Lesson 11: Connect the area model and the partial products method to the standard algorithm.

169

© 2018 Great Minds®. eureka-math.org

Name _____   Date _____

1. Solve the following expressions using the standard algorithm, the partial products method, and the area model.

a. 3 0 2 × 8

8(300 + 2)

(8 × _____) + (8 × _____)

b. 2 1 6 × 5

5 (_____ + _____ + _____)

(__ × _____) + (__ × _____) + (__ × _____)

c. 5 9 3 × 9

__(_____ + _____ + _____)

(__ × _____) + (__ × _____) + (__ × _____)

Lesson 11:   Connect the area model and the partial products method to the standard algorithm.

© 2018 Great Minds®. eureka-math.org

171

2. Solve using the partial products method.

   On Monday, 475 people visited the museum. On Saturday, there were 4 times as many visitors as there were on Monday. How many people visited the museum on Saturday?

3. Model with a tape diagram and solve.

   6 times as much as 384

Solve using the standard algorithm, the area model, the distributive property, or the partial products method.

4. $6,253 \times 3$

Lesson 11:   Connect the area model and the partial products method to the
                               standard algorithm.

© 2018 Great Minds®. eureka-math.org

EUREKA
MATH

5. 7 times as many as 3,073

6. A cafeteria makes 2,516 pounds of white rice and 608 pounds of brown rice every month. After 6 months, how many pounds of rice does the cafeteria make?

EUREKA MATH

Lesson 11: Connect the area model and the partial products method to the standard algorithm.

© 2018 Great Minds®. eureka-math.org

173

Name _____ Date _____

Use the RDW process to solve the following problems.

1. The table shows the number of stickers of various types in Chrissy's new sticker book. Chrissy's six friends each own the same sticker book. How many stickers do Chrissy and her six friends have altogether?

Type of Sticker	Number of Stickers
flowers	32
smiley faces	21
hearts	39

2. The small copier makes 437 copies each day. The large copier makes 4 times as many copies each day. How many copies does the large copier make each week?

3. Jared sold 194 Boy Scout chocolate bars. Matthew sold three times as many as Jared. Gary sold 297 fewer than Matthew. How many bars did Gary sell?

Lesson 12:   Solve two-step word problems, including multiplicative comparison.

177

© 2018 Great Minds®. eureka-math.org

4. a.   Write an equation that would allow someone to find the value of M.

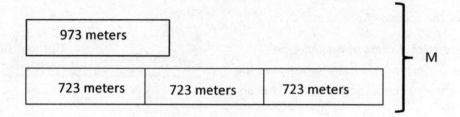

b.   Write your own word problem to correspond to the tape diagram, and then solve.

EUREKA
MATH®

Solve using the RDW process.

1. A banana costs 58¢. A pomegranate costs 3 times as much. What is the total cost of a pomegranate and 5 bananas?

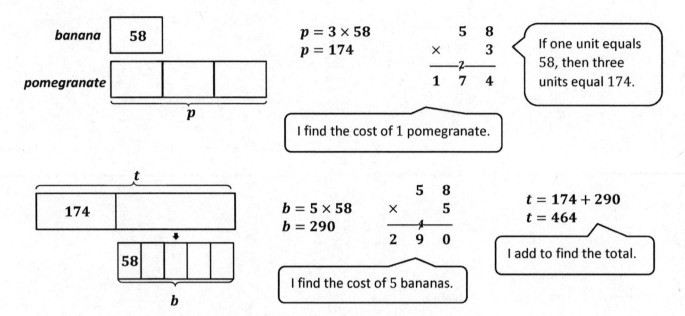

$p = 3 \times 58$
$p = 174$

If one unit equals 58, then three units equal 174.

I find the cost of 1 pomegranate.

$b = 5 \times 58$
$b = 290$

$t = 174 + 290$
$t = 464$

I add to find the total.

I find the cost of 5 bananas.

The total cost of a pomegranate and 5 bananas is 464¢.

2. Mr. Turner gave his 2 daughters $197 each. He gave his mother $325. He gave his wife money as well. If Mr. Turner gave a total of $3,000, how much did he give to his wife?

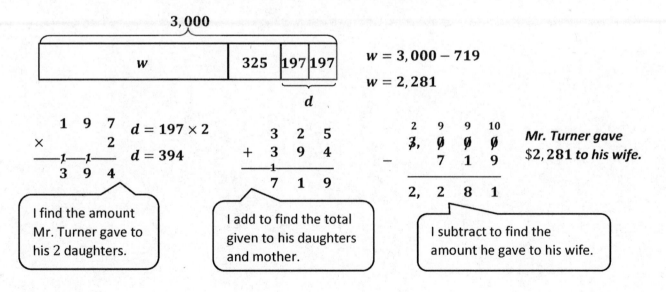

$w = 3,000 - 719$
$w = 2,281$

$d = 197 \times 2$
$d = 394$

I find the amount Mr. Turner gave to his 2 daughters.

I add to find the total given to his daughters and mother.

Mr. Turner gave $2,281 to his wife.

I subtract to find the amount he gave to his wife.

EUREKA MATH

Lesson 13: Use multiplication, addition, or subtraction to solve multi-step word problems.

© 2018 Great Minds®. eureka-math.org

179

Name _____ Date _____

Solve using the RDW process.

1. A pair of jeans costs $89. A jean jacket costs twice as much. What is the total cost of a jean jacket and 4 pairs of jeans?

2. Sarah bought a shirt on sale for $35. The original price of the shirt was 3 times that amount. Sarah also bought a pair of shoes on sale for $28. The original price of the shoes was 5 times that amount. Together, how much money did the shirt and shoes cost before they went on sale?

Lesson 13:   Use multiplication, addition, or subtraction to solve multi-step word problems.

© 2018 Great Minds®. eureka-math.org

181

3. All 3,000 seats in a theater are being replaced. So far, 5 sections of 136 seats and a sixth section containing 348 seats have been replaced. How many more seats do they still need to replace?

4. Computer Depot sold 762 reams of paper. Paper Palace sold 3 times as much paper as Computer Depot and 143 reams more than Office Supply Central. How many reams of paper were sold by all three stores combined?

Lesson 13: Use multiplication, addition, or subtraction to solve multi-step word problems.

EUREKA MATH

Use the RDW process to solve the following problems.

1. Marco has 19 tortillas. If he uses 2 tortillas for each quesadilla, what is the greatest number of quesadillas he can make? Will he have any extra tortillas? How many?

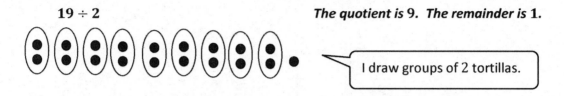

$19 \div 2$ *The quotient is 9. The remainder is 1.*

> I draw groups of 2 tortillas.

He can make up to 9 quesadillas. He will have 1 extra tortilla.

2. Coach Adam puts 31 players into teams of 8. How many teams does he make? If he makes a smaller team with the remaining players, how many players are on that team?

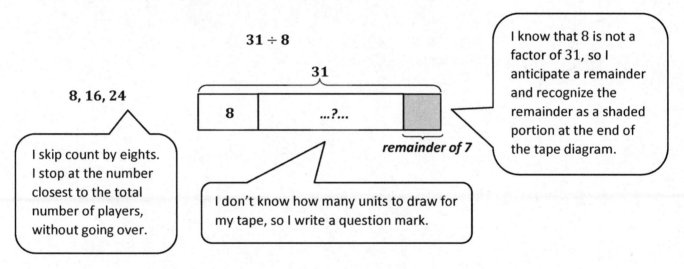

$31 \div 8$

31

8, 16, 24

8   ...?...

remainder of 7

> I know that 8 is not a factor of 31, so I anticipate a remainder and recognize the remainder as a shaded portion at the end of the tape diagram.

> I skip count by eights. I stop at the number closest to the total number of players, without going over.

> I don't know how many units to draw for my tape, so I write a question mark.

Coach Adam makes 3 teams. The smaller team has 7 players.

Name _____   Date _____

Use the RDW process to solve the following problems.

1. Linda makes booklets using 2 sheets of paper. She has 17 sheets of paper. How many of these booklets can she make? Will she have any extra paper? How many sheets?

2. Linda uses thread to sew the booklets together. She cuts 6 inches of thread for each booklet. How many booklets can she stitch with 50 inches of thread? Will she have any unused thread after stitching up the booklets? If so, how much?

3. Ms. Rochelle wants to put her 29 students into groups of 6. How many groups of 6 can she make? If she puts any remaining students in a smaller group, how many students will be in that group?

4. A trainer gives his horse, Caballo, 7 gallons of water every day from a 57-gallon container. How many days will Caballo receive his full portion of water from the container? On which number day will the trainer need to refill the container of water?

5. Meliza has 43 toy soldiers. She lines them up in rows of 5 to fight imaginary zombies. How many of these rows can she make? After making as many rows of 5 as she can, she puts the remaining soldiers in the last row. How many soldiers are in that row?

6. Seventy-eight students are separated into groups of 8 for a field trip. How many groups are there? The remaining students form a smaller group of how many students?

Lesson 14: Solve division word problems with remainders.

EUREKA MATH

Show division using an array.	Show division using an area model.

1. 21 ÷ 4

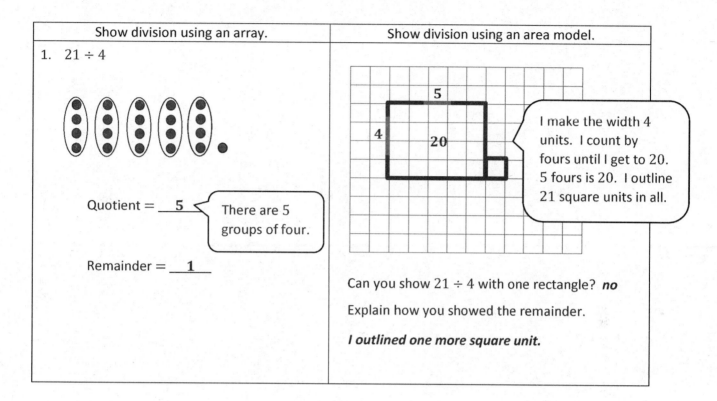

Quotient = __5__

> There are 5 groups of four.

Remainder = __1__

> I make the width 4 units. I count by fours until I get to 20. 5 fours is 20. I outline 21 square units in all.

Can you show 21 ÷ 4 with one rectangle? **no**

Explain how you showed the remainder.

I outlined one more square unit.

Solve using an array and area model.

2. 53 ÷ 7

> I can draw quickly without grid paper.

a. Array

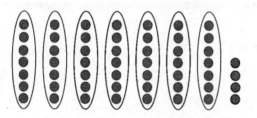

Quotient = 7   Remainder = 4

> The area model may be faster to draw, but no matter which model I use, I get the same answer!

b. Area Model

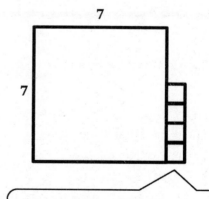

> I represent the remainder with 4 more square units.

Name _____ Date _____

Show division using an array.	Show division using an area model.
1. 24 ÷ 4	
Quotient = _____	
Remainder = _____	Can you show 24 ÷ 4 with one rectangle? _____
2. 25 ÷ 4	
Quotient = _____	Can you show 25 ÷ 4 with one rectangle? _____
Remainder = _____	Explain how you showed the remainder:

EUREKA MATH®

Lesson 15:   Understand and solve division problems with a remainder using the array and area models.

© 2018 Great Minds®. eureka-math.org

189

Solve using an array and area model. The first one is done for you.

Example: 25 ÷ 3

a.

Quotient = 8   Remainder = 1

b.

3. 44 ÷ 7

a.   b.

4. 34 ÷ 6

a.   b.

5. 37 ÷ 6

a.   b.

6. 46 ÷ 8

a.   b.

Lesson 15:   Understand and solve division problems with a remainder using the
               array and area models.

EUREKA
MATH

Show the division using disks. Relate your work on the place value chart to long division. Check your quotient and remainder by using multiplication and addition.

1. $9 \div 2$

> To model, the divisor represents the number of equal groups. The quotient represents the size of the groups.

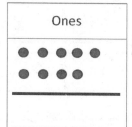

Ones

> I represent 9 ones, the whole, using place value disks.

> I make space on the chart to distribute the disks into 2 equal groups.

> 9 ones distributed evenly into 2 equal groups is 4 ones in each group. I cross them off as I distribute.

> 1 one remains because it cannot be distributed evenly into 2. I circle it to show it is a remainder.

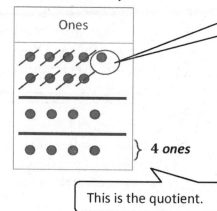

} **4 ones**

> This is the quotient.

$$\begin{array}{r} 4 \;\; R1 \\ 2\,\overline{)\,9} \\ -\,8 \\ \hline 1 \end{array}$$

quotient = __4__

remainder = __1__

> I check my division by multiplying the quotient times the divisor. I add the remainder. The sum is the whole.

Check your work.

$$\begin{array}{r} 4 \\ \times\,2 \\ \hline 8 \end{array} \qquad \begin{array}{r} 8 \\ +\,1 \\ \hline 9 \end{array}$$

EUREKA MATH

Lesson 16: Understand and solve two-digit dividend division problems with a remainder in the ones place by using place value disks.

191

© 2018 Great Minds®. eureka-math.org

2. $87 \div 4$

I represent the whole as 8 tens and 7 ones. I partition the chart into 4 equal groups below.

Tens	Ones

$8 \div 4 = 2$

8 tens distributed evenly among 4 groups is 2 tens.

$$\begin{array}{r} 2 \\ 4\,\overline{)\,8\ \ 7} \\ -\ \underline{8} \\ 0\ \ 7 \end{array}$$

$2 \times 4 = 8$

2 tens in each of the 4 groups is 8 tens.

$8 - 8 = 0$

We started with 8 tens and distributed 8 tens evenly. Zero tens and 7 ones remain in the whole.

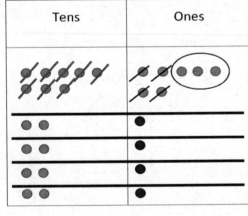

Tens	Ones

} 2 tens 1 one

$7 \div 4 = 1$

7 ones distributed evenly among 4 groups is 1 one.

$$\begin{array}{r} 2\ \ 1\ \ R3 \\ 4\,\overline{)\,8\ \ 7} \\ -\ \underline{8} \\ 0\ \ 7 \\ -\ \underline{4} \\ 3 \end{array}$$

$4 \times 1 = 4$

1 one in each of the 4 groups is 4 ones. Only 4 of the 7 ones were evenly distributed.

$7 - 4 = 3$

We started with 7 ones and distributed 4 ones evenly. 3 ones remain in the whole.

I record the remainder next to the quotient.

quotient = ___21___

remainder = ___3___

Check your work

$$\begin{array}{r} 2\ \ 1 \\ \times\quad\ 4 \\ \hline 8\ \ 4 \end{array} \qquad \begin{array}{r} 8\ \ 4 \\ +\quad\ 3 \\ \hline 8\ \ 7 \end{array}$$

EUREKA MATH

Name _____   Date _____

Show the division using disks. Relate your work on the place value chart to long division. Check your quotient and remainder by using multiplication and addition.

1. 7 ÷ 3

Ones

3 ⟌ 7

quotient = _____

remainder = _____

Check Your Work

$$\begin{array}{r} 2 \\ \times\ 3 \\ \hline \end{array}$$

2. 67 ÷ 3

Tens	Ones

3 ⟌ 6 7

quotient = _____

remainder = _____

Check Your Work

EUREKA MATH

Lesson 16:   Understand and solve two-digit dividend division problems with a remainder in the ones place by using place value disks.

193

© 2018 Great Minds®. eureka-math.org

3. 5 ÷ 2

Ones

2 | 5

quotient = _____

remainder = _____

Check Your Work

4. 85 ÷ 2

Tens	Ones

2 | 8 5

quotient = _____

remainder = _____

Check Your Work

Lesson 16: Understand and solve two-digit dividend division problems with a remainder in the ones place by using place value disks.

EUREKA MATH

5. $5 \div 4$

Ones

$4 \overline{)5}$

quotient = _____

remainder = _____

Check Your Work

6. $85 \div 4$

Tens	Ones

$4 \overline{)85}$

quotient = _____

remainder = _____

Check Your Work

Lesson 16: Understand and solve two-digit dividend division problems with a remainder in the ones place by using place value disks.

195

EUREKA MATH

© 2018 Great Minds®. eureka-math.org

Show the division using disks. Relate your model to long division. Check your quotient by using multiplication and addition.

1. $5 \div 4$

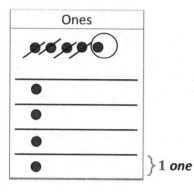

Ones

$}1$ *one*

$$\begin{array}{r} 1 \quad R1 \\ 4\overline{)5} \\ -\underline{4} \\ 1 \end{array}$$

quotient = __1__

remainder = __1__

Just like Lesson 16, I model the whole and partition the chart into 4 parts to represent the divisor.

Check your work.

$$\begin{array}{rr} 4 & 4 \\ \times\ 1 & +\ 1 \\ \hline 4 & 5 \end{array}$$

2. $53 \div 4$

After distributing 4 tens, 1 ten remains. I change 1 ten for 10 ones.

Now, I have 13 ones. I can distribute 12 ones evenly, but 1 one remains.

Tens	Ones

$}$ **1** *ten* **3** *ones*

$$\begin{array}{r} 1\ 3 \quad R1 \\ 4\overline{)5\ 3} \\ -\underline{4} \\ 1\ 3 \\ -\underline{1\ 2} \\ 1 \end{array}$$

quotient = __13__

remainder = __1__

Check your work.

$$\begin{array}{rr} 1\ 3 & 5\ 2 \\ \times\quad 4 & +\quad 1 \\ \hline 5\ 2 & 5\ 3 \end{array}$$

EUREKA MATH

Lesson 17:   Represent and solve division problems requiring decomposing a remainder in the tens.

197

© 2018 Great Minds®. eureka-math.org

Name _____   Date _____

Show the division using disks. Relate your model to long division. Check your quotient and remainder by using multiplication and addition.

1. 7 ÷ 2

Ones

2 ⟌ 7

Check Your Work

quotient = _____

remainder = _____

2. 73 ÷ 2

Tens	Ones

2 ⟌ 7 3

Check Your Work

quotient = _____

remainder = _____

Lesson 17:   Represent and solve division problems requiring decomposing a remainder in the tens.

199

EUREKA MATH

© 2018 Great Minds®. eureka-math.org

3. 6 ÷ 4

Ones

4 | 6

Check Your Work

quotient = _____

remainder = _____

4. 62 ÷ 4

Tens	Ones

4 | 6 2

Check Your Work

quotient = _____

remainder = _____

Lesson 17:   Represent and solve division problems requiring decomposing a remainder in the tens.

EUREKA MATH

5. 8 ÷ 3

Ones

3 | 8

Check Your Work

quotient = _____

remainder = _____

6. 84 ÷ 3

Tens	Ones

3 | 8 4

Check Your Work

quotient = _____

remainder = _____

EUREKA MATH

Lesson 17:   Represent and solve division problems requiring decomposing a remainder in the tens.

© 2018 Great Minds®. eureka-math.org

201

Solve using the standard algorithm. Check your quotient and remainder by using multiplication and addition.

1. $69 \div 3$

$$
\begin{array}{r}
2\ 3 \\
3\ \overline{)6\ 9} \\
-\ 6 \\
\hline
0\ 9 \\
-\ \ \ 9 \\
\hline
0
\end{array}
$$

$$
\begin{array}{r}
2\ 3 \\
\times\ \ \ 3 \\
\hline
6\ 9
\end{array}
$$

69 divided by 3 is 23.
And 23 times 3 is 69.

2. $57 \div 3$

I notice the divisor is the same in Problems 1 and 2. But the whole 69 is greater than the whole of 57. When the divisor is the same, the larger the whole, the larger the quotient.

$$
\begin{array}{r}
1\ 9 \\
3\ \overline{)5\ 7} \\
-\ 3 \\
\hline
2\ 7 \\
-\ 2\ 7 \\
\hline
0
\end{array}
$$

I distribute 3 tens. 2 tens remain. After decomposing, 20 ones plus 7 ones is 27 ones.

$$
\begin{array}{r}
1\ 9 \\
\times\ \ \ 3 \\
\hline
5\ 7
\end{array}
$$

When the wholes are nearly the same, the larger the divisor, the smaller the quotient. That's because the whole is divided into more equal groups.

3. $94 \div 5$

$$
\begin{array}{r}
1\ 8\ R4 \\
5\ \overline{)9\ 4} \\
-\ 5 \\
\hline
4\ 4 \\
-\ 4\ 0 \\
\hline
4
\end{array}
$$

$$
\begin{array}{r}
1\ 8 \\
\times\ \ \ 5 \\
\hline
9\ 0
\end{array}
$$

$$
\begin{array}{r}
9\ 0 \\
+\ \ \ 4 \\
\hline
9\ 4
\end{array}
$$

The quotient is 18 with a remainder of 4.

4. $97 \div 7$

$$
\begin{array}{r}
1\ 3\ R6 \\
7\ \overline{)9\ 7} \\
-\ 7 \\
\hline
2\ 7 \\
-\ 2\ 1 \\
\hline
6
\end{array}
$$

$$
\begin{array}{r}
1\ 3 \\
\times\ \ \ 7 \\
\hline
9\ 1
\end{array}
$$

$$
\begin{array}{r}
9\ 1 \\
+\ \ \ 6 \\
\hline
9\ 7
\end{array}
$$

I prove my division is correct by multiplying 13 by 7 and then adding 6 more.

7. $91 \div 6$

8. $91 \div 7$

9. $87 \div 3$

10. $87 \div 6$

11. $94 \div 8$

12. $94 \div 6$

Lesson 18:   Find whole number quotients and remainders.

EUREKA
MATH

1. Makhai says that $97 \div 3$ is 30 with a remainder of 7. He reasons this is correct because $(3 \times 30) + 7 = 97$. What mistake has Makhai made? Explain how he can correct his work.

   Makhai stopped dividing when he had 7 ones, but he can distribute them into 3 more groups of 2. If he does so, he can make 3 groups of 32 instead of just 30.

   > There are not enough ones to distribute into 3 groups. I record 1 one as the remainder.

   ```
           3 2  R1
        ┌───────
      3 │ 9 7
        − 9
        ───────
          0 7
        −   6
        ───────
            1
   ```

2. Four friends evenly share 52 dollars.

   a. They have 5 ten-dollar bills and 2 one-dollar bills. Draw a picture to show how the bills will be shared. Will they have to make change at any stage?

   > I unbundle a ten by drawing an arrow from the remaining 1 ten to 10 ones.

   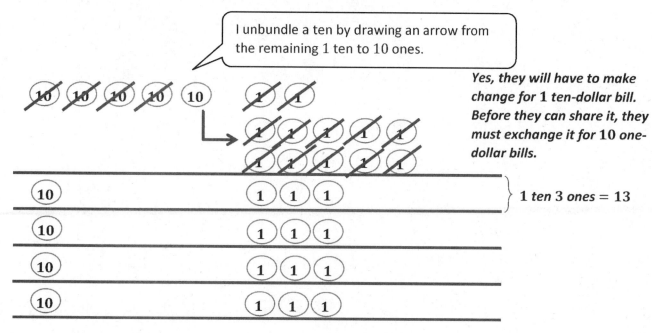

   Yes, they will have to make change for 1 ten-dollar bill. Before they can share it, they must exchange it for 10 one-dollar bills.

   1 ten 3 ones = 13

   b. Explain how they share the money evenly.

   Each friend gets 1 ten-dollar bill and 3 one-dollar bills.

3. Imagine you are writing a magazine article describing how to solve the problem $43 \div 3$ to new fourth graders. Write a draft to explain how you can keep dividing after getting a remainder of 1 ten in the first step.

Sample answer: This is how you divide 43 by 3. Think of it like 4 tens 3 ones divided into 3 groups. First, you want to distribute the tens. You can distribute 3 tens. Each group will have 1 ten. There will be 1 ten left over. That's okay. You can keep dividing. Just change 1 ten for 10 ones. Now you have 13 ones altogether. You can distribute 12 ones evenly. 3 groups of 4 ones is 12 ones. 1 one is remaining. So, your quotient is 14 R1. And that's how you divide 43 by 3.

```
        1  4  R1
    ┌─────────
  3 │ 4  3
    − 3
    ─────────
      1  3
    − 1  2
    ─────────
         1
```

Lesson 18: Explain remainders by using place value understanding and models.

EUREKA MATH

Name _____ Date _____

1. When you divide 86 by 4, there is a remainder of 2. Model this problem with place value disks. In the place value disk model, how can you see that there is a remainder?

2. Francine says that 86 ÷ 4 is 20 with a remainder of 6. She reasons this is correct because (4 × 20) + 6 = 86. What mistake has Francine made? Explain how she can correct her work.

3. The place value disk model is showing $67 \div 4$.
   Complete the model. Explain what happens to the
   2 tens that are remaining in the tens column.

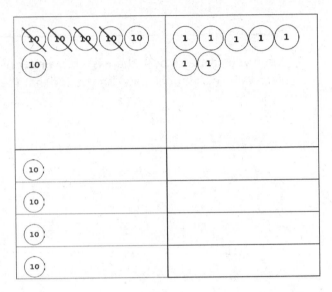

4. Two friends share 76 blueberries.

   a. To count the blueberries, they put them into small bowls of 10 blueberries. Draw a picture to show
      how the blueberries can be shared equally. Will they have to split apart any of the bowls
      of 10 blueberries when they share them?

   b. Explain how the friends can share the blueberries fairly.

Lesson 19:   Explain remainders by using place value understanding and models.

EUREKA
MATH

5. Imagine you are drawing a comic strip showing how to solve the problem 72 ÷ 4 to new fourth graders. Create a script to explain how you can keep dividing after getting a remainder of 3 tens in the first step.

Lesson 19:   Explain remainders by using place value understanding and models.

211

© 2018 Great Minds®. eureka-math.org

1. Paco solved a division problem by drawing an area model.

    a. Look at the area model. What division problem did Paco solve?

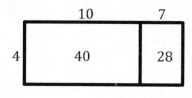

$$68 \div 4 = 17$$

I add the areas to find the whole. The width is the divisor. I add the two lengths to find the quotient.

    b. Show a number bond to represent Paco's area model. Start with the total, and then show how the total is split into two parts. Below the two parts, represent the total length using the distributive property, and then solve.

Dividing smaller numbers is easier for me than solving $68 \div 4$. I can solve mentally because these are easy facts.

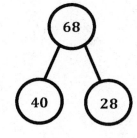

In the number bond, I record the whole (68) split into two parts (40 and 28).

$$(\underline{40} \div \underline{4}) + (\underline{28} \div \underline{4})$$
$$= \underline{10} + \underline{7}$$
$$= \underline{17}$$

2. Solve $76 \div 4$ using an area model. Explain the connection of the distributive property to the area model using words, pictures, or numbers.

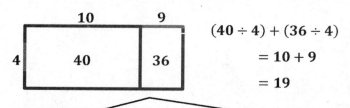

$$(40 \div 4) + (36 \div 4)$$
$$= 10 + 9$$
$$= 19$$

The area model is like a picture for the distributive model. Each rectangle represents a smaller division expression that we write in parentheses. The width of the rectangle is the divisor in each sentence. The two lengths are added together to get the quotient.

I think of 4 times how many lengths of ten get me close to 7 tens in the whole: 1 ten. Then, 4 times how many lengths of ones gets me close to the remaining 36 ones: 9 ones.

Name _____   Date _____

1. Maria solved a division problem by drawing an area model.

   a. Look at the area model. What division problem did Maria solve?

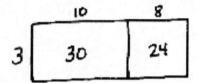

   b. Show a number bond to represent Maria's area model. Start with the total, and then show how the total is split into two parts. Below the two parts, represent the total length using the distributive property, and then solve.

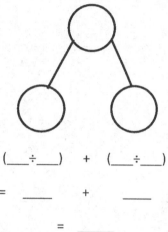

   (___÷___)   +   (___÷___)

   = _____   +   _____

   = _____

2. Solve 42 ÷ 3 using an area model. Draw a number bond, and use the distributive property to solve for the unknown length.

EUREKA
MATH

Lesson 20:   Solve division problems without remainders using the area model.

215

© 2018 Great Minds®. eureka-math.org

3. Solve 60 ÷ 4 using an area model. Draw a number bond to show how you partitioned the area, and represent the division with a written method.

4. Solve 72 ÷ 4 using an area model. Explain, using words, pictures, or numbers, the connection of the distributive property to the area model.

5. Solve 96 ÷ 6 using an area model and the standard algorithm.

EUREKA
MATH

1. Yahya solved the following division problem by drawing an area model.

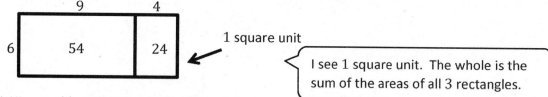

1 square unit

I see 1 square unit. The whole is the sum of the areas of all 3 rectangles.

   a. What division problem did he solve? **79 ÷ 6**

   b. Show how Yahya's model can be represented using the distributive property.

$$(54 \div 6) + (24 \div 6)$$
$$= 9 + 4$$
$$= 13$$

I remember to add a remainder of 1.

$$(6 \times 13) + 1 = 79$$

Solve the following problems using the area model. Support the area model with long division or the distributive property.

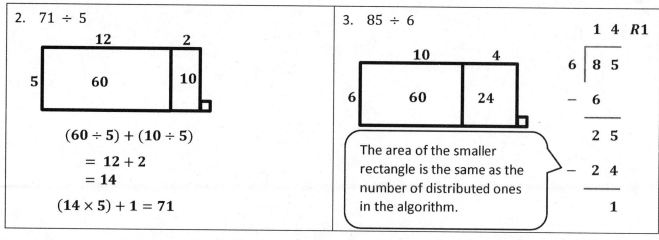

2. 71 ÷ 5

$$(60 \div 5) + (10 \div 5)$$
$$= 12 + 2$$
$$= 14$$
$$(14 \times 5) + 1 = 71$$

3. 85 ÷ 6

The area of the smaller rectangle is the same as the number of distributed ones in the algorithm.

```
      1 4 R1
  6 | 8 5
    - 6
    -----
      2 5
    - 2 4
    -----
        1
```

4. Eighty-nine marbles were placed equally in 4 bags. How many marbles were in each bag? How many marbles are left over?

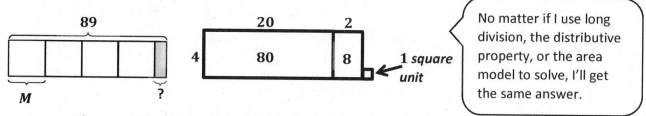

1 square unit

No matter if I use long division, the distributive property, or the area model to solve, I'll get the same answer.

There are 22 marbles in each bag. 1 marble is left over.

Name _____   Date _____

1. Solve 35 ÷ 2 using an area model. Use long division and the distributive property to record your work.

2. Solve 79 ÷ 3 using an area model. Use long division and the distributive property to record your work.

3. Paulina solved the following division problem by drawing an area model.

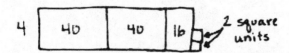

   a. What division problem did she solve?

   b. Show how Paulina's model can be represented using the distributive property.

Solve the following problems using the area model. Support the area model with long division or the distributive property.

4. 42 ÷ 3	5. 43 ÷ 3
6. 52 ÷ 4	7. 54 ÷ 4
8. 61 ÷ 5	9. 73 ÷ 3

Lesson 21:   Solve division problems with remainders using the area model.

EUREKA MATH

10. Ninety-seven lunch trays were placed equally in 4 stacks. How many lunch trays were in each stack? How many lunch trays will be left over?

Lesson 21: Solve division problems with remainders using the area model.

221

© 2018 Great Minds®. eureka-math.org

1. Record the factors of the given numbers as multiplication sentences and as a list in order from least to greatest. Classify each as prime (P) or composite (C).

	Multiplication Sentences	Factors	P or C
a.	5 $1 \times 5 = 5$	The factors of 5 are **1, 5**	**P**
b.	18 $1 \times 18 = 18$ $2 \times 9 = 18$ $3 \times 6 = 18$	The factors of 18 are **1, 2, 3, 6, 9, 18**	**C**

> I know a number is prime if it has only two factors. I know a number is composite if it has more than two factors.

2. Find all factors for the following number, and classify the number as prime or composite. Explain your classification of prime or composite.

Factor Pairs for 12	
1	12
2	6
3	4

12 is composite. I know that it is composite because it has more than two factors.

> I think of the multiplication facts that have a product of 12.

3. Jenny has 25 beads to divide evenly among 4 friends. She thinks there will be no leftovers. Use what you know about factor pairs to explain whether or not Jenny is correct.

Jenny is not correct. There will be leftovers. I know this because if 4 is one of the factors, there is no whole number that multiplies by 4 to get 25 as a product. There will be one bead left over.

> $4 \times 6 = 24$ and $4 \times 7 = 28$. There is no factor pair for 4 that results in a product of 25.

Lesson 22:   Find factor pairs for numbers to 100, and use understanding of factors to define prime and composite.

Name _____ Date _____

1. Record the factors of the given numbers as multiplication sentences and as a list in order from least to greatest. Classify each as prime (P) or composite (C). The first problem is done for you.

	Multiplication Sentences	Factors	P or C
a.	8 $1 \times 8 = 8$ $2 \times 4 = 8$	The factors of 8 are: 1, 2, 4, 8	C
b.	10	The factors of 10 are:	
c.	11	The factors of 11 are:	
d.	14	The factors of 14 are:	
e.	17	The factors of 17 are:	
f.	20	The factors of 20 are:	
g.	22	The factors of 22 are:	
h.	23	The factors of 23 are:	
i.	25	The factors of 25 are:	
j.	26	The factors of 26 are:	
k.	27	The factors of 27 are:	
l.	28	The factors of 28 are:	

EUREKA MATH®

Lesson 22: Find factor pairs for numbers to 100, and use understanding of factors to define prime and composite.

225

2. Find all factors for the following numbers, and classify each number as prime or composite. Explain your classification of each as prime or composite.

Factor Pairs for 19		Factor Pairs for 21		Factor Pairs for 24	

3. Bryan says that only even numbers are composite.

    a. List all of the odd numbers less than 20 in numerical order.

    b. Use your list to show that Bryan's claim is false.

4. Julie has 27 grapes to divide evenly among 3 friends. She thinks there will be no leftovers. Use what you know about factor pairs to explain whether or not Julie is correct.

Lesson 22:   Find factor pairs for numbers to 100, and use understanding of factors to define prime and composite.

© 2018 Great Minds®. eureka-math.org

EUREKA MATH

1. Explain your thinking, or use division to answer the following.

Is 2 a factor of 96?	Is 3 a factor of 96?
Yes. *96 is an even number. 2 is a factor of every even number.*	$\begin{array}{r} 3\ 2 \\ 3\overline{\smash{)}9\ 6} \\ \underline{-\ 9} \\ 0\ 6 \\ \underline{-\ \ \ 6} \\ 0 \end{array}$ *Yes, 3 is a factor of 96. When I divide 96 by 3, my answer is 32.*
Is 4 a factor of 96?	Is 5 a factor of 96?
$\begin{array}{r} 2\ 4 \\ 4\overline{\smash{)}9\ 6} \\ \underline{-\ 8} \\ 1\ 6 \\ \underline{-\ 1\ 6} \\ 0 \end{array}$ *Yes, 4 is a factor of 96. When I divide 96 by 4, my answer is 24.*	*No, 5 is not a factor of 96. 96 does not have a 5 or 0 in the ones place. All numbers that have a 5 as a factor have a 5 or 0 in the ones place.*

> I use what I know about factors to solve. Thinking about whether 2 is a factor or 5 is a factor is easy. Threes and fours are harder, so I divide to see if they are factors. 96 is divisible by both 3 and 4, so they are both factors of 96.

2. Use the associative property to find more factors of 28 and 32.

a. $28 = 14 \times 2$

$= (\underline{\ 7\ } \times 2) \times 2$

$= \underline{\ 7\ } \times (2 \times 2)$

$= \underline{\ 7\ } \times 4$

$= \underline{\ 28\ }$

b. $32 = \underline{\ 8\ } \times 2$

$= (\underline{\ 2\ } \times 4) \times 4$

$= \underline{\ 2\ } \times (4 \times 4)$

$= \underline{\ 2\ } \times 16$

$= \underline{\ 32\ }$

> I find more factors of the whole number by breaking down one of the factors into smaller parts and then associating the factors differently using parentheses.

EUREKA MATH

Lesson 23: Use division and the associative property to test for factors and observe patterns.

© 2018 Great Minds®. eureka-math.org

227

3. In class, we used the associative property to show that when 6 is a factor, then 2 and 3 are factors, because $6 = 2 \times 3$. Use the fact that $12 = 2 \times 6$ to show that 2 and 6 are factors of 36, 48, and 60.

$$36 = 12 \times 3$$
$$= (2 \times 6) \times 3$$
$$= 2 \times (6 \times 3)$$
$$= 2 \times 18$$
$$= 36$$

$$48 = 12 \times 4$$
$$= (2 \times 6) \times 4$$
$$= 2 \times (6 \times 4)$$
$$= 2 \times 24$$
$$= 48$$

$$60 = 12 \times 5$$
$$= (2 \times 6) \times 5$$
$$= 2 \times (6 \times 5)$$
$$= 2 \times 30$$
$$= 60$$

I rewrite the number sentences, substituting 2×6 for 12. I can move the parentheses because of the associative property and then solve. This helps to show that both 2 and 6 are factors of 36, 48, and 60.

4. The first statement is false. The second statement is true. Explain why using words, pictures, or numbers.

   If a number has 2 and 8 as factors, then it has 16 as a factor.
   If a number has 16 as a factor, then both 2 and 8 are factors.

The first statement is false. For example, 8 has both 2 and 8 as factors, but it does not have 16 as a factor. The second statement is true. Any number that can be divided exactly by 16 can also be divided by 2 and 8 instead since $16 = 2 \times 8$. Example: $2 \times 16 = 32$

$$2 \times (2 \times 8) = 32$$

I give examples to help with my explanation.

Use division and the associative property to test for factors and observe patterns.

EUREKA MATH

Name _____   Date _____

1. Explain your thinking or use division to answer the following.

a. Is 2 a factor of 72?	b. Is 2 a factor of 73?
c. Is 3 a factor of 72?	d. Is 2 a factor of 60?
e. Is 6 a factor of 72?	f. Is 4 a factor of 60?
g. Is 5 a factor of 72?	h. Is 8 a factor of 60?

Lesson 23:   Use division and the associative property to test for factors and
observe patterns.

2. Use the associative property to find more factors of 12 and 30.

a. $12 = 6 \times 2$

$= (\_\_\_ \times 2) \times 2$

$= \_\_\_ \times (2 \times 2)$

$= \_\_\_ \times \_\_\_$

$= \_\_\_$

b. $30 = \_\_\_ \times 5$

$= (\_\_\_ \times 3) \times 5$

$= \_\_\_ \times (3 \times 5)$

$= \_\_\_ \times 15$

$= \_\_\_$

3. In class, we used the associative property to show that when 6 is a factor, then 2 and 3 are factors, because $6 = 2 \times 3$. Use the fact that $10 = 5 \times 2$ to show that 2 and 5 are factors of 70, 80, and 90.

$70 = 10 \times 7$ $80 = 10 \times 8$ $90 = 10 \times 9$

4. The first statement is false. The second statement is true. Explain why, using words, pictures, or numbers.

If a number has 2 and 6 as factors, then it has 12 as a factor.
If a number has 12 as a factor, then both 2 and 6 are factors.

Lesson 23:   Use division and the associative property to test for factors and observe patterns.

EUREKA
MATH

1. Write the multiples of 3 starting from 36. Time yourself for 1 minute. See how many multiples you can write.

   36, 39, 42, 45, 48, 51, 54, 57, 60, 63, 66, 69, 72, 75, 78, 81, 84, 87, 90, 93, 96, 99, 102, 105, 108, 111, 114

   > I skip-count by threes starting with 36.

2. List the numbers that have 28 as a multiple.

   1, 2, 4, 7, 14, 28

   > This is just like finding the factor pairs of a number. If I say "28" when I skip-count by a number, that means 28 is a multiple of that number.

3. Use mental math, division, or the associative property to solve.

   a. Is 15 a multiple of 3? __yes__   Is 3 a factor of 15? __yes__

   > $3 \times 5 = 15$, so 3 is a factor of 15.

   b. Is 34 a multiple of 6? __no__   Is 6 a factor of 34? __no__

   Is 32 a multiple of 8? __yes__   Is 32 a factor of 8? __no__

   > If a number is a multiple of another number, it means that, when I skip-count, I say that number.

   > 8 is a factor of 32, but 32 is not a factor of 8.

Lesson 24: Determine if a whole number is a multiple of another number.

231

4. Follow the directions below.

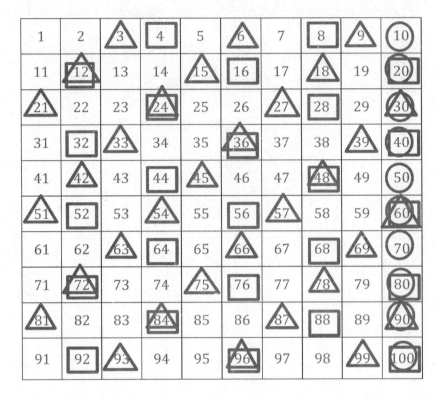

a. Circle the multiples of 10. When a number is a multiple of 10, what do you notice about the number in the ones place?

   When a number is a multiple of 10, the number in the ones place is always a zero.

b. Draw a square around the multiples of 4. When a number is a multiple of 4, what are the possible numbers in the ones digit?

   When a number is a multiple of 4, the possible number in the ones digit is 2, 4, 6, 8, or 0.

c. Put a triangle on the multiples of 3. Choose one. What do you notice about the sum of the digits? Choose another one. What do you notice about the sum of the digits?

   15 → *The sum of the digits is 6.*

   75 → *The sum of the digits is 12.*

   > If I look at more multiples of 3, I see that the sum of their digits is 3, 6, 9, 12, 15, or 18. Each of those numbers is a multiple of 3.

  Lesson 24: Determine if a whole number is a multiple of another number.

EUREKA MATH

Name _____ Date _____

1. For each of the following, time yourself for 1 minute. See how many multiples you can write.

   a. Write the multiples of 5 starting from 75.

   b. Write the multiples of 4 starting from 40.

   c. Write the multiples of 6 starting from 24.

2. List the numbers that have 30 as a multiple.

3. Use mental math, division, or the associative property to solve. (Use scratch paper if you like.)

   a. Is 12 a multiple of 3? _____   Is 3 a factor of 12? _____

   b. Is 48 a multiple of 8? _____   Is 48 a factor of 8? _____

   c. Is 56 a multiple of 6? _____   Is 6 a factor of 56? _____

4. Can a prime number be a multiple of any other number except itself? Explain why or why not.

5. Follow the directions below.

1	2	3	4	5	6	7	8	9	10
11	12	13	14	15	16	17	18	19	20
21	22	23	24	25	26	27	28	29	30
31	32	33	34	35	36	37	38	39	40
41	42	43	44	45	46	47	48	49	50
51	52	53	54	55	56	57	58	59	60
61	62	63	64	65	66	67	68	69	70
71	72	73	74	75	76	77	78	79	80
81	82	83	84	85	86	87	88	89	90
91	92	93	94	95	96	97	98	99	100

a. Underline the multiples of 6. When a number is a multiple of 6, what are the possible values for the ones digit?

b. Draw a square around the multiples of 4. Look at the multiples of 4 that have an odd number in the tens place. What values do they have in the ones place?

c. Look at the multiples of 4 that have an even number in the tens place. What values do they have in the ones place? Do you think this pattern would continue with multiples of 4 that are larger than 100?

d. Circle the multiples of 9. Choose one. What do you notice about the sum of the digits? Choose another one. What do you notice about the sum of the digits?

Lesson 24:   Determine if a whole number is a multiple of another number.

1. Follow the directions.

    Shade the number 1.

    a. Circle the first unmarked number.

    b. Cross off every multiple of that number except the one you circled. If it's already crossed off, skip it.

    c. Repeat Steps (a) and (b) until every number is either circled or crossed off.

    d. Shade every crossed out number.

1	2	3	4	5	6	7	8	9	10
11	12	13	14	15	16	17	18	19	20
21	22	23	24	25	26	27	28	29	30
31	32	33	34	35	36	37	38	39	40
41	42	43	44	45	46	47	48	49	50
51	52	53	54	55	56	57	58	59	60
61	62	63	64	65	66	67	68	69	70
71	72	73	74	75	76	77	78	79	80
81	82	83	84	85	86	87	88	89	90
91	92	93	94	95	96	97	98	99	100

I cross off every multiple of 2 except for the number 2.

Name _____   Date _____

1. A student used the sieve of Eratosthenes to find all prime numbers less than 100. Create a step-by-step set of directions to show how it was completed. Use the word bank to help guide your thinking as you write the directions. Some words may be used just once, more than once, or not at all.

Word Bank

composite	cross out
number	shade
circle	X
multiple	prime

Directions for completing the sieve of Eratosthenes activity:

EUREKA MATH

Lesson 25:   Explore properties of prime and composite numbers to 100 by using multiples.

© 2018 Great Minds®. eureka-math.org

239

2. What do all of the numbers that are crossed out have in common?

3. What do all of the circled numbers have in common?

4. There is one number that is neither crossed out nor circled. Why is it treated differently?

Lesson 25: Explore properties of prime and composite numbers to 100 by using multiples.

EUREKA MATH®

1. Draw place value disks to represent the following problems. Rewrite each in unit form and solve.

a. $80 \div 4 = $ __20__

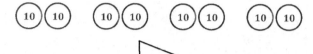

     8 tens $\div 4 = $ __2 tens__

> 2 tens is the same as 20.

> I distribute 8 tens into 4 groups. There are 2 tens in each group.

b. $800 \div 4 = $ __200__

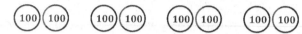

   __8 hundreds__ $\div 4 = $ __2 hundreds__

> I think of 800 in unit form as 8 hundreds.

> 8 hundreds divided equally into 4 groups is 2 hundreds.

c. $150 \div 3 = $ __50__

   __15 tens__ $\div 3 = $ __5 tens__

> I think of 150 as 1 hundred 5 tens, but that doesn't help me to divide because I can't partition a hundreds disk into 3 equal groups. To help me to divide, I think of 150 as 15 tens.

d. $1,500 \div 3 = $ __500__

   __15 hundreds__ $\div 3 = $ __5 hundreds__

> This is just like the last problem except the unit is hundreds instead of tens.

2. Solve for the quotient. Rewrite each in unit form.

a. $900 \div 3 = \mathbf{300}$	b. $140 \div 2 = \mathbf{70}$	c. $1,500 \div 5 = \mathbf{300}$	d. $200 \div 5 = \mathbf{40}$
9 hundreds ÷ 3 *= 3 hundreds*	*14 tens ÷ 2* *= 7 tens*	*15 hundreds ÷ 5* *= 3 hundreds*	*20 tens ÷ 5* *= 4 tens*

These problems are very similar to what I just did. The difference is that I do not draw disks. I rewrite the numbers in unit form to help me solve.

3. An ice cream shop sold $2,800 of ice cream in August, which was 4 times as much as was sold in May. How much ice cream was sold at the ice cream shop in May?

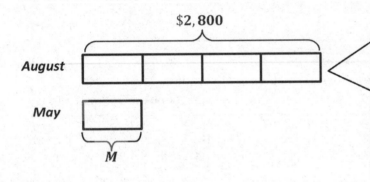

$2,800

August

May

M

I draw a tape diagram to show the ice cream sales for the month of August and the month of May. The tape for August is 4 times as long as the tape for May. 2,800 in unit form is 28 hundreds. If 4 units is 28 hundreds, 1 unit must be 28 hundreds ÷ 4. Since May is equal to 1 unit, the ice cream sales for May was $700.

28 hundreds ÷ 4 = 7 hundreds

$700 *of ice cream was sold at the ice cream shop in May.*

Lesson 26: Divide multiples of 10, 100, and 1,000 by single-digit numbers.

EUREKA MATH

Name _____   Date _____

1. Draw place value disks to represent the following problems. Rewrite each in unit form and solve.

   a. $6 \div 3 =$ _____

       6 ones $\div 3 =$ _____ ones

   b. $60 \div 3 =$ _____

       6 tens $\div 3 =$ _____

   c. $600 \div 3 =$ _____

       _____ $\div 3 =$ _____

   d. $6{,}000 \div 3 =$ _____

       _____ $\div 3 =$ _____

2. Draw place value disks to represent each problem. Rewrite each in unit form and solve.

   a. $12 \div 4 =$ _____

       12 ones $\div 4 =$ _____ ones

   b. $120 \div 4 =$ _____

       _____ $\div 4 =$ _____

   c. $1{,}200 \div 4 =$ _____

       _____ $\div 4 =$ _____

3. Solve for the quotient. Rewrite each in unit form.

a. 800 ÷ 4 = 200 8 hundreds ÷ 4 = 2 hundreds	b. 900 ÷ 3 = _____	c. 400 ÷ 2 = _____	d. 300 ÷ 3 = _____
e. 200 ÷ 4 = _____ 20 tens ÷ 4 = ____ tens	f. 160 ÷ 2 = _____	g. 400 ÷ 5 = _____	h. 300 ÷ 5 = _____
i. 1,200 ÷ 3 = _____ 12 hundreds ÷ 3 = ____ hundreds	j. 1,600 ÷ 4 = _____	k. 2,400 ÷ 4 = _____	l. 3,000 ÷ 5 = _____

4. A fleet of 5 fire engines carries a total of 20,000 liters of water. If each truck holds the same amount of water, how many liters of water does each truck carry?

Lesson 26:   Divide multiples of 10, 100, and 1,000 by single-digit numbers.

EUREKA
MATH

5. Jamie drank 4 times as much juice as Brodie. Jamie drank 280 milliliters of juice. How much juice did Brodie drink?

6. A diner sold $2,400 worth of French fries in June, which was 4 times as much as was sold in May. How many dollars' worth of French fries were sold at the diner in May?

EUREKA
MATH

Lesson 26:   Divide multiples of 10, 100, and 1,000 by single-digit numbers.

245

© 2018 Great Minds®. eureka-math.org

Divide. Model using place value disks, and record using the algorithm.

$426 \div 3$

hundreds	tens	ones
● ● ● ●	● ●	● ● ● ● ● ●

> I represent 426 as 4 hundreds 2 tens 6 ones.

> I make space on the chart to distribute the disks into 3 equal groups.

hundreds	tens	ones
⫽ ⫽ ⫽ ⊘	● ●	● ● ● ● ● ●
●		
●		
●		

> I remember from Lesson 16 to divide starting in the largest unit.

> 4 hundreds divided by 3 is 1 hundred.

$$\begin{array}{r} 1 \\ 3\overline{\smash{)}426} \\ -\underline{3} \\ 1 \end{array}$$

> 1 hundred in each group times 3 groups is 3 hundreds.

> We started with 4 hundreds and evenly divided 3 hundreds. 1 hundred remains, which I've circled.

EUREKA MATH

Lesson 27: Represent and solve division problems with up to a three-digit dividend numerically and with place value disks requiring decomposing a remainder in the hundreds place.

© 2018 Great Minds®. eureka-math.org

247

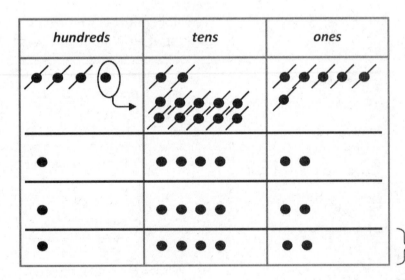

hundreds	tens	ones

$$
\begin{array}{r}
1 \\
3\,\overline{)4\ \ 2\ \ 6} \\
-\ \underline{3} \\
1\ \ 2
\end{array}
$$

> I remember from Lesson 17 that when there are remaining units that can't be divided, I decompose them as 10 of the next smallest unit. So 1 hundred is decomposed as 10 tens. Now there are 12 tens to divide.

hundreds	tens	ones

> I continue to distribute tens and ones, and I record each step of the algorithm.

$$
\begin{array}{r}
1\ \ 4\ \ 2 \\
3\,\overline{)4\ \ 2\ \ 6} \\
-\ \underline{3} \\
1\ \ 2 \\
-\ \underline{1\ \ 2} \\
0\ \ 6 \\
-\ \underline{6} \\
0
\end{array}
$$

1 *hundred* **4** *tens* **2** *ones*

> The value in each group equals the quotient.

Lesson 27:   Represent and solve division problems with up to a three-digit dividend numerically and with place value disks requiring decomposing a remainder in the hundreds place.

© 2018 Great Minds®. eureka-math.org

EUREKA MATH®

Name _____   Date _____

1. Divide. Use place value disks to model each problem.

a. 346 ÷ 2

b. 528 ÷ 2

Lesson 27:   Represent and solve division problems with up to a three-digit
               dividend numerically and with place value disks requiring
               decomposing a remainder in the hundreds place.

© 2018 Great Minds®. eureka-math.org

249

c.   516 ÷ 3

d.   729 ÷ 3

Lesson 27:   Represent and solve division problems with up to a three-digit dividend numerically and with place value disks requiring decomposing a remainder in the hundreds place.

2. Model using place value disks, and record using the algorithm.

a. 648 ÷ 4

Disks   Algorithm

b. 755 ÷ 5

Disks   Algorithm

c. 964 ÷ 4

Disks   Algorithm

Lesson 27: Represent and solve division problems with up to a three-digit dividend numerically and with place value disks requiring decomposing a remainder in the hundreds place.

© 2018 Great Minds®. eureka-math.org

251

1. Divide. Check your work by multiplying. Draw disks on a place value chart as needed.

a. $217 \div 4$

hundreds	tens	ones

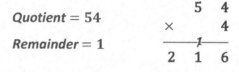

Quotient = 54

Remainder = 1

$$\begin{array}{r} 5\ 4 \\ \times \quad\ 4 \\ \hline 2\ 1\ 6 \end{array}$$

$$\begin{array}{r} 2\ 1\ 6 \\ +\quad\ \ 1 \\ \hline 2\ 1\ 7 \end{array}$$

} 5 tens 4 ones

I can't distribute 2 hundreds evenly among the 4 groups. I decompose each hundred as 10 tens. Now I have 21 tens.

I check my answer by multiplying the quotient and the divisor, and then I add the remainder. My answer of 217 matches the whole in the division expression.

b. $743 \div 3$

$$\begin{array}{r} 2\ 4\ 7\ \text{R2} \\ 3\overline{\smash{)}7\ 4\ 3} \\ -\ 6 \\ \hline 1\ 4 \\ -\ 1\ 2 \\ \hline 2\ 3 \\ -\ 2\ 1 \\ \hline 2 \end{array}$$

$$\begin{array}{r} 2\ 4\ 7 \\ \times \quad\ 3 \\ \hline 7\ 4\ 1 \end{array} \qquad \begin{array}{r} 7\ 4\ 1 \\ +\quad\ \ 2 \\ \hline 7\ 4\ 3 \end{array}$$

I visualize each step on the place value chart as I record the steps of the algorithm.

Lesson 28:   Represent and solve three-digit dividend division with divisors of 2, 3, 4, and 5 numerically.

253

EUREKA MATH

2. Constance ran 620 meters around the 4 sides of a square field. How many meters long was each side of the field?

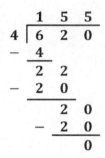

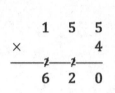

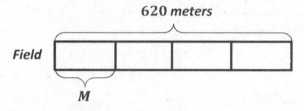

Each side of the field was 155 meters.

Lesson 28:   Represent and solve three-digit dividend division with divisors of 2, 3, 4, and 5 numerically.

© 2018 Great Minds®. eureka-math.org

EUREKA
MATH®

Name _____   Date _____

1. Divide. Check your work by multiplying. Draw disks on a place value chart as needed.

a. 378 ÷ 2

b. 795 ÷ 3

c. 512 ÷ 4

Lesson 28:   Represent and solve three-digit dividend division with divisors
of 2, 3, 4, and 5 numerically.

© 2018 Great Minds®. eureka-math.org

255

d. 492 ÷ 4

e. 539 ÷ 3

f. 862 ÷ 5

Lesson 28:   Represent and solve three-digit dividend division with divisors
of 2, 3, 4, and 5 numerically.

© 2018 Great Minds®. eureka-math.org

EUREKA
MATH

g. 498 ÷ 3

h. 783 ÷ 5

i. 621 ÷ 4

Lesson 28: Represent and solve three-digit dividend division with divisors of 2, 3, 4, and 5 numerically.

257

EUREKA MATH®

j.   531 ÷ 4

2   Selena's dog completed an obstacle course that was 932 meters long. There were 4 parts to the course, all equal in length. How long was 1 part of the course?

Represent and solve three-digit dividend division with divisors
of 2, 3, 4, and 5 numerically.

© 2018 Great Minds®. eureka-math.org

EUREKA
MATH

1. Divide, and then check using multiplication.

   3,268 ÷ 4

   ```
          8  1  7
     4 | 3, 2  6  8
       -  3  2
          ──────
             0  6
          -     4
                ──────
                2  8
             -  2  8
                ──────
                   0
   ```

   > I divide just as I learned to in Lessons 16, 17, 27, and 28. The challenge now is that the whole is larger, so I record the steps of the algorithm using long division and not using the place value chart.

   ```
          8  1  7
     ×           4
     ─────────────
     3,  2  6  8
   ```

   > I check the answer by multiplying the quotient and the divisor. The product is equal to the whole.

2. A school buys 3 boxes of pencils. Each box has an equal number of pencils. There are 4,272 pencils altogether. How many pencils are in 2 boxes?

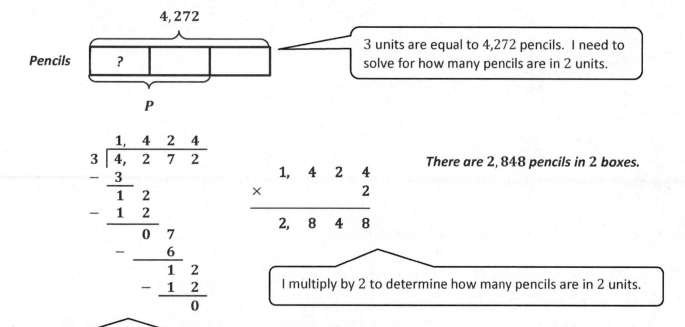

   Pencils

   4,272

   | ? | | |

   P

   > 3 units are equal to 4,272 pencils. I need to solve for how many pencils are in 2 units.

   ```
          1, 4  2  4
     3 | 4, 2  7  2
       -  3
          ──────
          1  2
       -  1  2
          ──────
             0  7
          -     6
                ──────
                1  2
             -  1  2
                ──────
                   0
   ```

   ```
        1, 4  2  4
     ×           2
     ──────────────
        2, 8  4  8
   ```

   There are 2,848 pencils in 2 boxes.

   > I multiply by 2 to determine how many pencils are in 2 units.

   > I find how many pencils are in 1 unit by dividing 4 272 by 3. There are 1 424 pencils in 1 unit.

EUREKA MATH

Lesson 29: Represent numerically four-digit dividend division with divisors of 2, 3, 4, and 5, decomposing a remainder up to three times.

259

© 2018 Great Minds®. eureka-math.org

Name _____ Date _____

1. Divide, and then check using multiplication.

---

a. 2,464 ÷ 4

---

b. 1,848 ÷ 3

---

c. 9,426 ÷ 3

Lesson 29: Represent numerically four-digit dividend division with divisors of 2, 3, 4, and 5, decomposing a remainder up to three times.

261

© 2018 Great Minds®. eureka-math.org

d. 6,587 ÷ 2

e. 5,445 ÷ 3

f. 5,425 ÷ 2

Lesson 29:   Represent numerically four-digit dividend division with divisors
of 2, 3, 4, and 5, decomposing a remainder up to three times.

© 2018 Great Minds®. eureka-math.org

EUREKA
MATH

g. 8,467 ÷ 3

h. 8,456 ÷ 3

i. 4,937 ÷ 4

Lesson 29:   Represent numerically four-digit dividend division with divisors of 2, 3, 4, and 5, decomposing a remainder up to three times.

© 2018 Great Minds®. eureka-math.org

263

j. $6,173 \div 5$

2.   A truck has 4 crates of apples. Each crate has an equal number of apples. Altogether, the truck is carrying 1,728 apples. How many apples are in 3 crates?

Represent numerically four-digit dividend division with divisors of 2, 3, 4, and 5, decomposing a remainder up to three times.

EUREKA
MATH

Divide. Check your solutions by multiplying.

1. 705 ÷ 2

```
      3  5  2   R1
   2 | 7  0  5
     - 6
       1  0
     - 1  0
          0  5
           - 4
             1
```

I decompose 1 hundred as 10 tens. There are no other tens to distribute. So I keep dividing, this time in the tens.

Once I divide the 10 tens, there are no tens remaining. But I must keep dividing. There are still 5 ones to divide.

```
      3  5  2
   ×         2
   ‾‾‾‾‾‾‾‾‾‾‾
      7  0  4
```

```
      7  0  4
   +         1
   ‾‾‾‾‾‾‾‾‾‾‾
      7  0  5
```

2. 6,250 ÷ 5

```
       1  2  5  0
   5 | 6, 2  5  0
     - 5
       1  2
     - 1  0
          2  5
        - 2  5
             0  0
           -    0
                0
```

This time when I divide, there are no ones to distribute. 0 ones divided by 5 is 0 ones. I place a 0 in the ones place of the quotient to show that there are no ones.

```
      1, 2  5  0
   ×            5
   ‾‾‾‾‾‾‾‾‾‾‾‾‾‾
      6, 2  5  0
```

3. 3,220 ÷ 4

```
          8  0  5
   4 | 3, 2  2  0
     - 3  2
          0  2
        -    0
             2  0
           - 2  0
                0
```

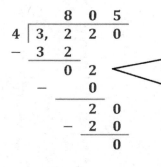

2 tens can't be evenly divided by 4, so I record 0 tens in the quotient. But I must continue the steps of the algorithm: 0 tens times 4 equals 0 tens. 2 tens minus 0 tens is 2 tens.

```
         8  0  5
   ×           4
   ‾‾‾‾‾‾‾‾‾‾‾‾‾
      3, 2  2  0
```

EUREKA MATH   **Lesson 30:**   Solve division problems with a zero in the dividend or with a zero in   **265**
                                    the quotient.

© 2018 Great Minds®. eureka-math.org

Name _____ Date _____

Divide. Check your solutions by multiplying.

1. 409 ÷ 5

2. 503 ÷ 2

3. 831 ÷ 4

4. 602 ÷ 3

Lesson 30: Solve division problems with a zero in the dividend or with a zero in the quotient.

© 2018 Great Minds®. eureka-math.org

267

5. 720 ÷ 3

6. 6,250 ÷ 5

7. 2,060 ÷ 5

8. 9,031 ÷ 2

Lesson 30:   Solve division problems with a zero in the dividend or with a zero in
the quotient.

EUREKA
MATH®

9. 6,218 ÷ 4

10. 8,000 ÷ 4

Lesson 30: Solve division problems with a zero in the dividend or with a zero in the quotient.

269

Solve the following problems. Draw tape diagrams to help you solve. Identify if the group size or the number of groups is unknown.

1. 700 liters of water was shared equally among 4 aquariums. How many liters of water does each aquarium have?

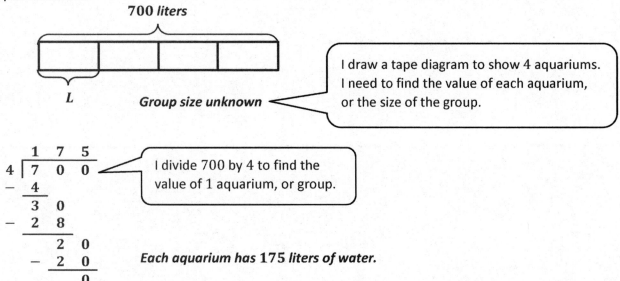

700 *liters*

L   ***Group size unknown***

I draw a tape diagram to show 4 aquariums. I need to find the value of each aquarium, or the size of the group.

```
      1  7  5
   4 | 7  0  0
   -  4
      3  0
   -  2  8
         2  0
      -  2  0
            0
```

I divide 700 by 4 to find the value of 1 aquarium, or group.

*Each aquarium has **175** liters of water.*

2. Emma separated 824 donuts into boxes. Each box contained 4 donuts. How many boxes of donuts did Emma fill?

824

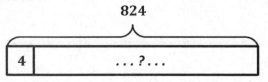

4   *. . . ? . . .*

Number of groups unknown

I do not know how many boxes were filled. I show one group of 4. I draw three dots, a question mark, and three dots to indicate that the groups of 4 continue. The number of groups is unknown.

```
      2  0  6
   4 | 8  2  4
   -  8
      0  2
      -  0
         2  4
      -  2  4
            0
```

*Emma filled **206** boxes of donuts.*

I divide 824 by 4 to find the number of groups.

EUREKA MATH   Lesson 31:   Interpret division word problems as either *number of groups unknown*   **271**
or *group size unknown*.

© 2018 Great Minds®. eureka-math.org

Name _____ Date _____

Solve the following problems. Draw tape diagrams to help you solve. Identify if the group size or the number of groups is unknown.

1. 500 milliliters of juice was shared equally by 4 children. How many milliliters of juice did each child get?

2. Kelly separated 618 cookies into baggies. Each baggie contained 3 cookies. How many baggies of cookies did Kelly make?

3. Jeff biked the same distance each day for 5 days. If he traveled 350 miles altogether, how many miles did he travel each day?

Lesson 31: Interpret division word problems as either *number of groups unknown* or *group size unknown*.

273

© 2018 Great Minds®. eureka-math.org

4. A piece of ribbon 876 inches long was cut by a machine into 4-inch long strips to be made into bows. How many strips were cut?

5. Five Martians equally share 1,940 Groblarx fruits. How many Groblarx fruits will 3 of the Martians receive?

Lesson 31:   Interpret division word problems as either *number of groups unknown* or *group size unknown*.

© 2018 Great Minds®. eureka-math.org

EUREKA MATH

Solve the following problems. Draw tape diagrams to help you solve. If there is a remainder, shade in a small portion of the tape diagram to represent that portion of the whole.

1. The clown has 1,649 balloons. It takes 8 balloons to make a balloon animal. How many balloon animals can the clown make?

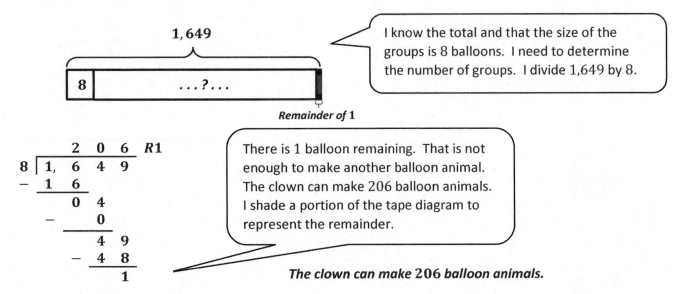

I know the total and that the size of the groups is 8 balloons. I need to determine the number of groups. I divide 1,649 by 8.

There is 1 balloon remaining. That is not enough to make another balloon animal. The clown can make 206 balloon animals. I shade a portion of the tape diagram to represent the remainder.

The clown can make 206 balloon animals.

2. In 7 days, Cassidy threw a total of 609 pitches. If she threw the same number of pitches each day, how many pitches did she throw in one day?

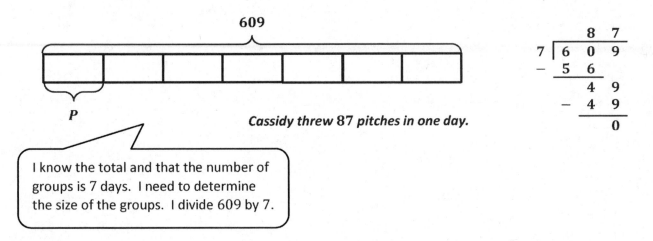

Cassidy threw 87 pitches in one day.

I know the total and that the number of groups is 7 days. I need to determine the size of the groups. I divide 609 by 7.

EUREKA
MATH®

Lesson 32:   Interpret and find whole number quotients and remainders to solve
one-step division word problems with larger divisors of 6, 7, 8, and 9.

275

© 2018 Great Minds®. eureka-math.org

Name _____   Date _____

Solve the following problems. Draw tape diagrams to help you solve. If there is a remainder, shade in a small portion of the tape diagram to represent that portion of the whole.

1. Meneca bought a package of 435 party favors to give to the guests at her birthday party. She calculated that she could give 9 party favors to each guest. How many guests is she expecting?

2. 4,000 pencils were donated to an elementary school. If 8 classrooms shared the pencils equally, how many pencils did each class receive?

3. 2,008 kilograms of potatoes were packed into sacks weighing 8 kilograms each. How many sacks were packed?

Lesson 32:   Interpret and find whole number quotients and remainders to solve
one-step division word problems with larger divisors of 6, 7, 8, and 9.

277

4. A baker made 7 batches of muffins. There was a total of 252 muffins. If there was the same number of muffins in each batch, how many muffins were in a batch?

5. Samantha ran 3,003 meters in 7 days. If she ran the same distance each day, how far did Samantha run in 3 days?

Lesson 32: Interpret and find whole number quotients and remainders to solve one-step division word problems with larger divisors of 6, 7, 8, and 9.

EUREKA MATH

1. Tyler solved a division problem by drawing this area model.

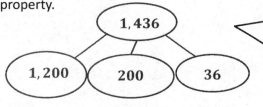

	300	50	9
4	1,200	200	36

> The total area is $1,200 + 200 + 36 = 1,436$. The width is 4. The length is $300 + 50 + 9 = 359$. $A \div w = l$.

a. What division problem did he solve?

   Tyler solved $1,436 \div 4 = 359$.

b. Show a number bond to represent Tyler's area model, and represent the total length using the distributive property.

1,436
1,200   200   36

> My number bond shows the same whole and parts as the area model. To represent the length, I divide each of the smaller areas by the width of 4.

$(1,200 \div 4) + (200 \div 4) + (36 \div 4)$

$= \quad 300 \quad + \quad 50 \quad + \quad 9$

$= \quad 359$

> I decompose the area of 591 into smaller parts that are easy to divide by 3. I start with the hundreds. I distribute 3 hundreds. The area remaining to distribute is 291. I distribute 27 tens. The area remaining to distribute is 21 ones. I distribute the ones. I have a side length of $100 + 90 + 7 = 197$.

2.

a. Draw an area model to solve $591 \div 3$.

	100	90	7
3	300	270	21

$591 \div 3 = 197$

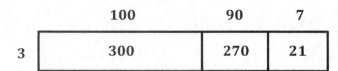

> 3 hundreds, 27 tens, and 21 ones are all multiples of 3, which is the width and divisor.

EUREKA
MATH

Lesson 33:   Explain the connection of the area model of division to the long
              division algorithm for three- and four-digit dividends.   279

© 2018 Great Minds®. eureka-math.org

b.   Draw a number bond to represent this problem.

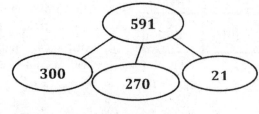

$$(300 \div 3) + (270 \div 3) + (21 \div 3)$$

$$= \quad 100 \ + \ 90 \ + \ 7$$

$$= \quad 197$$

> My number bond shows the same whole and parts as the area model. To represent the length, I divide each of the smaller areas by the width of 3. I get $100 + 90 + 7 = 197$.

c.   Record your work using the long division algorithm.

```
        1  9  7
     3| 5  9  1
      - 3
        ‾‾‾
        2  9
      - 2  7
        ‾‾‾‾‾
           2  1
         - 2  1
           ‾‾‾‾‾
              0
```

Lesson 33:   Explain the connection of the area model of division to the long
division algorithm for three- and four-digit dividends.

© 2018 Great Minds®. eureka-math.org

EUREKA
MATH

Name _____   Date _____

1. Arabelle solved the following division problem by drawing an area model.

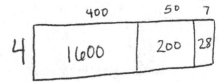

    a. What division problem did she solve?

    b. Show a number bond to represent Arabelle's area model, and represent the total length using the distributive property.

_____

2. a. Solve 816 ÷ 4 using the area model. There is no remainder in this problem.

    b. Draw a number bond and use a written method to record your work from Part (a).

Lesson 33:   Explain the connection of the area model of division to the long   281
               division algorithm for three- and four-digit dividends.

© 2018 Great Minds®. eureka-math.org

3. a. Draw an area model to solve 549 ÷ 3.

b. Draw a number bond to represent this problem.

c. Record your work using the long division algorithm.

---

4. a. Draw an area model to solve 2,762 ÷ 2.

b. Draw a number bond to represent this problem.

c. Record your work using the long division algorithm.

Lesson 33:   Explain the connection of the area model of division to the long division algorithm for three- and four-digit dividends.

EUREKA MATH

1. Use the associative property to rewrite each expression. Solve using disks, and then complete the number sentences.

I rename 30 as (3 × 10), and then I group the factor of 10 with 27.

I draw 2 tens 7 ones. I show 10 times as many by shifting the disks one place to the left.

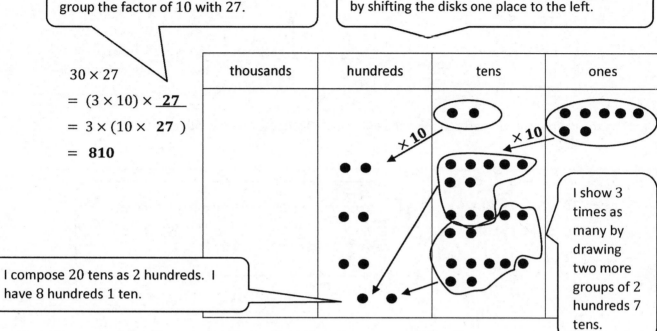

30×27

$= (3 \times 10) \times \underline{\ 27\ }$

$= 3 \times (10 \times \underline{\ 27\ })$

$= \mathbf{810}$

I show 3 times as many by drawing two more groups of 2 hundreds 7 tens.

I compose 20 tens as 2 hundreds. I have 8 hundreds 1 ten.

2. Use the associative property and place value disks to solve.

thousands	hundreds	tens	ones

20×28

$= (2 \times 10) \times 28$

$= 2 \times (10 \times 28)$

$= \mathbf{560}$

By decomposing 20 into 2 and 10, I think about the product being twice as much as 28 tens.

3. Use the associative property without place value disks to solve.

60×54

$= (6 \times 10) \times 54$

$= 6 \times (10 \times 54)$

$= 3{,}240$

$$\begin{array}{r} 5 \quad 4 \quad 0 \\ \times \qquad\qquad 6 \\ \hline {}_{1} \\ 3{,} \quad 2 \quad 4 \quad 0 \end{array}$$

> I rename 60 as 6×10. Ten times as many as 54 ones is 54 tens. I multiply 6 times 540.

4. Use the distributive property to solve the following. Distribute the second factor.

40×56

$= (40 \times 50) + (40 \times 6)$

$= 2{,}000 + 240$

$= 2{,}240$

> I use unit language to help me solve mentally. Four tens times 5 tens is 20 hundreds. And 4 tens times 6 ones is 24 tens.

Lesson 34:   Multiply two-digit multiples of 10 by two-digit numbers using a place value chart.

EUREKA
MATH

Name _____ Date _____

1. Use the associative property to rewrite each expression. Solve using disks, and then complete the number sentences.

    a.   20 × 34

         =(_____ × 10) × 34

         = ___ × (10 × 34)

         =_____

hundreds	tens	ones

    b. 30 × 34

         = (3 × 10) ×_____

         = 3 × (10 × ___)

         =_____

thousands	hundreds	tens	ones

    c. 30 × 42

         = (3 ×_____) × _____

         = 3 × (10 × _____)

         = _____

thousands	hundreds	tens	ones

2. Use the associative property and place value disks to solve.

    a. 20×16 b. 40×32

3. Use the associative property without place value disks to solve.

    a. 30×21 b. 60×42

4. Use the distributive property to solve the following. Distribute the second factor.

    a. 40×43 b. 70×23

EUREKA
MATH

1. Use an area model to represent the following expression. Then, record the partial products vertically and solve.

40×27

I write 40 as the width and decompose 27 as 20 and 7 for the length.

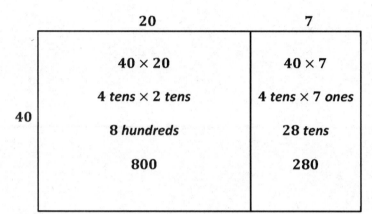

	20	**7**
40	40×20 *4 tens × 2 tens* *8 hundreds* **800**	40×7 *4 tens × 7 ones* *28 tens* **280**

I solve for each of the smaller areas.

```
        2   7
    ×   4   0
   _____
        2   8   0
    +   8   0   0
   _____
   1,   0   8   0
```

I record the partial products. The partial products have the same value as the areas of the smaller rectangles.

2. Visualize the area model, and solve the following expression numerically.

30×66

```
            6   6
        ×   3   0
       _____
            1   8   0
    +   1,  8   0   0
       _____
        1,  9   8   0
```

To solve, I visualize the area model. I see the width as 30 and the length as $60 + 6$. 3 tens × 6 ones = 18 tens. 3 tens × 6 tens = 18 hundreds. I record the partial products. I find the total. $180 + 1{,}800 = 1{,}980$.

EUREKA MATH

Lesson 35: Multiply two-digit multiples of 10 by two-digit numbers using the area model.

287

© 2018 Great Minds®. eureka-math.org

Name _____   Date _____

Use an area model to represent the following expressions. Then, record the partial products and solve.

1. 30 × 17

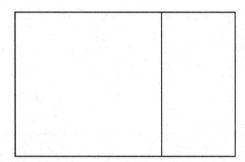

```
      1 7
   ×  3 0
   _____

+  _____
   _____
```

2. 40 × 58

```
      5 8
   ×  4 0
   _____

+  _____
   _____
```

3. 50 × 38

```
      3 8
   ×  5 0
   _____

+  _____
   _____
```

EUREKA
MATH

Lesson 35: Multiply two-digit multiples of 10 by two-digit numbers using the area model.

© 2018 Great Minds®. eureka-math.org

289

Draw an area model to represent the following expressions. Then, record the partial products vertically and solve.

4. 60×19

5. 20×44

Visualize the area model, and solve the following expressions numerically.

6. 20×88

7. 30×88

8. 70×47

9. 80×65

Lesson 35: Multiply two-digit multiples of 10 by two-digit numbers using the area model.

EUREKA MATH

1.

   a. In each of the two models pictured below, write the expressions that determine the area of each of the four smaller rectangles.

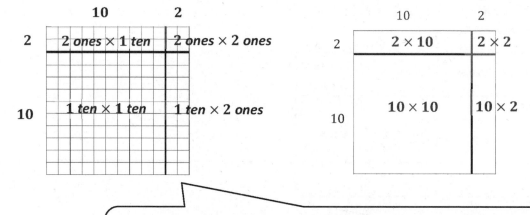

> I write the expressions that determine the area of each of the four smaller rectangles. The area of each smaller rectangle is equal to its width times its length. I can write the expressions in unit form or standard form.

   b. Using the distributive property, rewrite the area of the large rectangle as the sum of the areas of the four smaller rectangles. Express the area first in number form and then read it in unit form.

$$12 \times 12 = (2 \times \underline{2}) + (2 \times \underline{10}) + (10 \times \underline{2}) + (10 \times \underline{10})$$

> I write the expressions of the areas of the four smaller rectangles. I use the area models to help me. I say, "12 × 12 = (2 ones × 2 ones) + (2 ones × 1 ten) + (1 ten × 2 ones) + (1 ten × 1 ten)."

2. Use an area model to represent the following expression. Record the partial products vertically and solve.

15×33

	30	**3**
5	5 *ones* × 3 *tens*	5 *ones* × 3 *ones*
10	1 *ten* × 3 *tens*	1 *ten* × 3 *ones*

```
        3  3
   ×    1  5
   ─────────
        1  5
     1  5  0
        3  0
 +   3  0  0
   ─────────
     4  9  5
```

> I write the expressions that represent the areas of the four smaller rectangles. I record each partial product vertically. I find the sum of the areas of the four smaller rectangles.

3. Visualize the area model, and solve the following numerically using four partial products. (You may sketch an area model if it helps.)

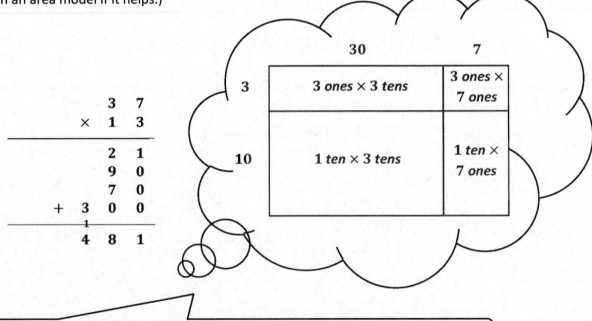

```
        3  7
   ×    1  3
   ─────────
        2  1
        9  0
        7  0
 +   3  0  0
       1
   ─────────
     4  8  1
```

	30	**7**
3	3 *ones* × 3 *tens*	3 *ones* × 7 *ones*
10	1 *ten* × 3 *tens*	1 *ten* × 7 *ones*

> To solve, I visualize the area model. I record the partial products. I find the total.

EUREKA MATH®

Name _____   Date _____

1. a.   In each of the two models pictured below, write the expressions that determine the area of each of the four smaller rectangles.

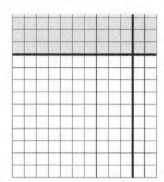

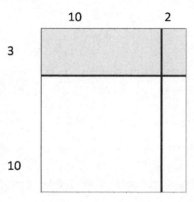

b. Using the distributive property, rewrite the area of the large rectangle as the sum of the areas of the four smaller rectangles. Express first in number form, and then read in unit form.

13 × 12 = (3 × _____) + (3 × _____) + (10 × _____) + (10 × _____)

Use an area model to represent the following expression. Record the partial products and solve.

2. 17 × 34

```
          3 4
      ×   1 7
      _____

      _____

      _____
  +
      _____
```

EUREKA MATH

Lesson 36:   Multiply two-digit by two-digit numbers using four partial products.

293

© 2018 Great Minds®. eureka-math.org

Draw an area model to represent the following expressions. Record the partial products vertically and solve.

3. 45×18 4. 45×19

Visualize the area model and solve the following numerically using four partial products. (You may sketch an area model if it helps.)

5. 12×47 6. 23×93

7. 23×11 8. 23×22

Lesson 36:   Multiply two-digit by two-digit numbers using four partial products.

EUREKA
MATH

1. Solve 37 × 54 using 4 partial products and 2 partial products. Remember to think in terms of units as you solve. Write an expression to find the area of each smaller rectangle in the area model. Match each partial product to its area on the models.

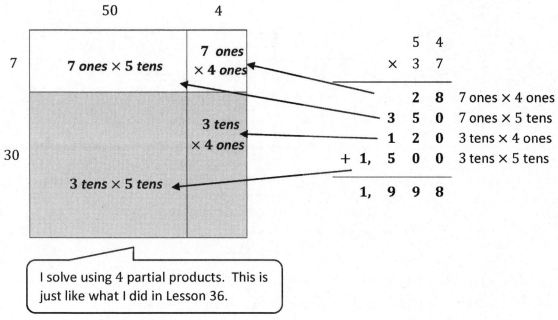

I solve using 4 partial products. This is just like what I did in Lesson 36.

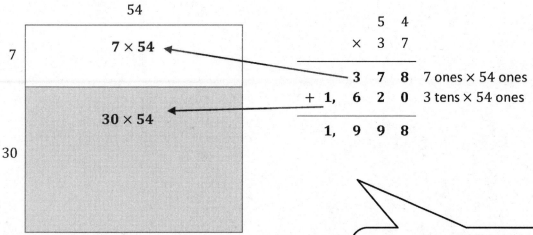

To show 2 partial products, I combine the values of the top two rectangles, and I combine the values of the bottom two rectangles.

I know one partial product is represented by the white portion of the large rectangle. The other partial product is represented by the shaded portion.

EUREKA MATH

Lesson 37: Transition from four partial products to the standard algorithm for two-digit by two-digit multiplication.

295

© 2018 Great Minds®. eureka-math.org

2. Solve 38 × 46 using 2 partial products and an area model. Match each partial product to its area on the model.

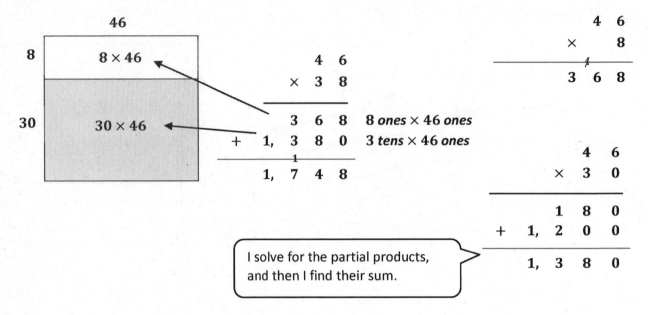

```
        4  6
     ×  3  8
     ─────────
        3  6  8      8 ones × 46 ones
  +  1, 3  8  0      3 tens × 46 ones
        1
     ─────────
     1, 7  4  8
```

```
           4  6
        ×     8
     ─────────
           4
        3  6  8
```

```
           4  6
        ×  3  0
     ─────────
           1  8  0
  +     1, 2  0  0
     ─────────
        1, 3  8  0
```

> I solve for the partial products, and then I find their sum.

3. Solve the following using 2 partial products. Visualize the area model to help you.

```
         7  4
      ×  2  5
   ─────────
         3  7  0      5 × 74
   + 1,  4  8  0      20 × 74
         1
   ─────────
     1,  8  5  0
```

```
           7  4
        ×     5
     ─────────
           2
        3  7  0
```

```
           7  4
        ×  2  0
     ─────────
           8  0
  +     1, 4  0  0
     ─────────
        1, 4  8  0
```

> I visualize the 2 partial products of 5 ones × 74 and 2 tens × 74. I solve for the partial products and then find their sum.

   Lesson 37:   Transition from four partial products to the standard algorithm for two-digit by two-digit multiplication.

© 2018 Great Minds®. eureka-math.org

EUREKA MATH®

Name _____   Date _____

1. Solve 26 × 34 using 4 partial products and 2 partial products. Remember to think in terms of units as you
    solve. Write an expression to find the area of each smaller rectangle in the area model.

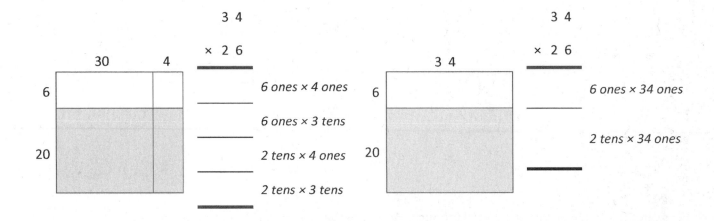

2. Solve using 4 partial products and 2 partial products. Remember to think in terms of units as you solve.
    Write an expression to find the area of each smaller rectangle in the area model.

EUREKA
MATH

Lesson 37:   Transition from four partial products to the standard algorithm for
two-digit by two-digit multiplication.

297

© 2018 Great Minds®. eureka-math.org

3. Solve 52 × 26 using 2 partial products and an area model. Match each partial product to its area on the model.

4. Solve the following using 2 partial products. Visualize the area model to help you.

a.
```
      6 8
  ×   2 3
  ─────────
```
_____ × _____

_____

_____ × _____

─────────

b.
```
      4 9
  ×   3 3
  ─────────
```
_____ × _____

_____

_____ × _____

─────────

c.
```
      1 6
  ×   2 5
  ─────────
```

d.
```
      5 4
  ×   7 1
  ─────────
```

Lesson 37: Transition from four partial products to the standard algorithm for two-digit by two-digit multiplication.

EUREKA MATH

1. Express 38×53 as two partial products using the distributive property. Solve.

```
        53
   ┌──────────┐
 8 │  8 × 53  │
   ├──────────┤
   │          │
30 │  30 × 53 │
   │          │
   └──────────┘
```

$38 \times 53 = (\underline{\textbf{8}} \text{ fifty-threes}) + (\underline{\textbf{30}} \text{ fifty-threes})$

```
              5   3
         ×    3   8
        ─────────────
              4   2   4     8  × 53
       +  1,  5   9   0    30  × 53
          1   1
        ─────────────
          2,  0   1   4
```

> I can solve for each of the partial products and find their sum to verify that I solved the 2-digit by 2-digit algorithm correctly.

```
              5   3                      5   3
        ×         8                ×     3   0
       ──────────────              ──────────────
          4   2   4                      9   0
                                  +  1,  5   0   0
                                  ──────────────
                                     1,  5   9   0
```

2. Express 34×44 as two partial products using the distributive property. Solve.

```
        44
   ┌──────────┐
 4 │  4 × 44  │
   ├──────────┤
   │          │
30 │  30 × 44 │
   │          │
   └──────────┘
```

$34 \times 44 = (\underline{\textbf{4}} \times \underline{\textbf{44}}) + (\underline{\textbf{30}} \times \underline{\textbf{44}})$

```
              4   4
        ×     3   4
       ──────────────
              1   7   6     4  × 44
       +  1,  3   2   0    30  × 44
       ──────────────
          1,  4   9   6
```

```
          4   4                      4   4
        ×     4                ×     3   0
       ──────────              ──────────────
          1   7   6                  1   2   0
                              +  1,  2   0   0
                              ──────────────
                                 1,  3   2   0
```

EUREKA MATH   **Lesson 38:**   Transition from four partial products to the standard algorithm for two-digit by two-digit multiplication.   **299**

© 2018 Great Minds®. eureka-math.org

3. Solve the following using two partial products.

$$
\begin{array}{r}
6\ \ 2 \\
\times\quad 4\ \ 3 \\
\hline
1\ \ 8\ \ 6 \\
+\quad 2,\ 4\ \ 8\ \ 0 \\
\hline
2,\ 6\ \ 6\ \ 6
\end{array}
$$

$\underline{\ 3\ } \times \underline{\ 62\ }$

$\underline{40} \times \underline{\ 62\ }$

I think of 3 sixty-twos + 40 sixty-twos.

4. Solve using the multiplication algorithm.

   62×36

$$
\begin{array}{r}
3\ \ 6 \\
\times\quad 6\ \ 2 \\
\hline
7\ \ 2 \\
+\quad 2,\ 1\ \ 6\ \ 0 \\
\hline
2,\ 2\ \ 3\ \ 2
\end{array}
$$

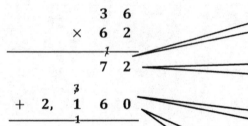

2 ones × 6 ones = 12 ones. I represent 12 ones as 1 ten 2 ones.

2 ones × 3 tens = 6 tens. 6 tens + 1 ten = 7 tens. I cross off 1 ten to show that I add it to 6 tens.

6 tens × 6 ones = 36 tens. I represent 36 tens as 3 hundreds 6 tens 0 ones.

6 tens × 3 tens = 18 hundreds.
18 hundreds + 3 hundreds = 21 hundreds.
I cross off 3 hundreds to show that I add it to 18 hundreds.

Lesson 38: Transition from four partial products to the standard algorithm for two-digit by two-digit multiplication.

© 2018 Great Minds®. eureka-math.org

EUREKA
MATH

Name _____   Date _____

1. Express 26 × 43 as two partial products using the distributive property. Solve.

43

6

20

26 × 43 = (_____ forty-threes) + (_____ forty-threes)

    4 3
×   2 6
_____

    6 × _____
_____

   20 × _____
_____

2. Express 47 × 63 as two partial products using the distributive property. Solve.

63

7

40

47 × 63 = (_____ sixty-threes) + (_____ sixty-threes)

    6 3
×   4 7
_____

   _____ × _____
_____

   _____ × _____
_____

3. Express 54 × 67 as two partial products using the distributive property. Solve.

54 × 67 = (___ × _____) + (___ × _____)

    6 7
×   5 4
_____

   _____ × _____
_____

   _____ × _____
_____

EUREKA MATH

Lesson 38:   Transition from four partial products to the standard algorithm for
             two-digit by two-digit multiplication.

301

© 2018 Great Minds®. eureka-math.org

4. Solve the following using two partial products.

```
      5 2
  ×   3 4
  _____
  _____   _____ × _____
  _____
  _____   _____ × _____
  _____
```

5. Solve using the multiplication algorithm.

```
      8 6
  ×   5 6
  _____
  _____   _____ × _____
  _____
  _____   _____ × _____
  _____
```

6. 54 × 52

7. 44 × 76

Lesson 38:   Transition from four partial products to the standard algorithm for two-digit by two-digit multiplication.

© 2018 Great Minds®. eureka-math.org

EUREKA
MATH

8. 63 × 63

9. 68 × 79

Lesson 38: Transition from four partial products to the standard algorithm for two-digit by two-digit multiplication.

303

Grade 4
Module 4

1. Use the following directions to draw a figure in the box below.

    a. Draw two points: *J* and *K*.

    b. Use a straightedge to draw $\overleftrightarrow{JK}$. ── I read this as "line *JK*."

    c. Draw a new point that is on $\overleftrightarrow{JK}$. Label it *L*.

    d. Draw a point not on $\overleftrightarrow{JK}$. Label it *M*.

    e. Construct $\overline{LM}$. ── I read this as "line segment *LM*."

    f. Use the points you've already labeled to name two angles.   ∠*JLM*, ∠*MLK*

    g. Identify the angles you've labeled by drawing an arc to indicate the position of the angles.

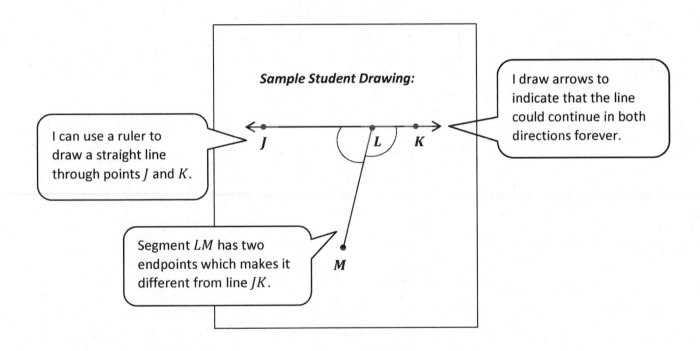

Sample Student Drawing:

I draw arrows to indicate that the line could continue in both directions forever.

I can use a ruler to draw a straight line through points *J* and *K*.

Segment *LM* has two endpoints which makes it different from line *JK*.

EUREKA MATH   Lesson 1:   Identify and draw points, lines, line segments, rays, and angles.   Recognize them in various contexts and familiar figures.   **307**

© 2018 Great Minds®. eureka-math.org

2.

a. Observe the familiar figures below. Label some points on each figure.

b. Use those points to label and name representations of each of the following in the table below: ray, line, line segment, and angle. Extend segments to show lines and rays.

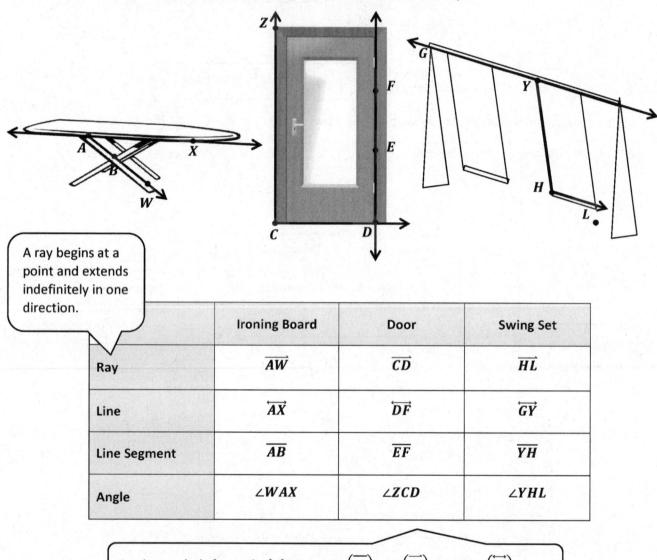

A ray begins at a point and extends indefinitely in one direction.

	Ironing Board	Door	Swing Set
Ray	$\overrightarrow{AW}$	$\overrightarrow{CD}$	$\overrightarrow{HL}$
Line	$\overleftrightarrow{AX}$	$\overleftrightarrow{DF}$	$\overleftrightarrow{GY}$
Line Segment	$\overline{AB}$	$\overline{EF}$	$\overline{YH}$
Angle	$\angle WAX$	$\angle ZCD$	$\angle YHL$

I write symbols for angle ($\angle$), segment ($\frown$), ray ($\frown$), and line ($\overleftrightarrow{}$).

Identify and draw points, lines, line segments, rays, and angles. Recognize them in various contexts and familiar figures.

EUREKA MATH

Name _____   Date _____

1. Use the following directions to draw a figure in the box to the right.

   a. Draw two points: W and X.

   b. Use a straightedge to draw $\overrightarrow{WX}$.

   c. Draw a new point that is not on $\overrightarrow{WX}$. Label it Y.

   d. Draw $\overline{WY}$.

   e. Draw a point not on $\overrightarrow{WX}$ or $\overline{WY}$. Call it Z.

   f. Construct $\overleftrightarrow{YZ}$.

   g. Use the points you've already labeled to name one angle. _____

2. Use the following directions to draw a figure in the box to the right.

   a. Draw two points: W and X.

   b. Use a straightedge to draw $\overline{WX}$.

   c. Draw a new point that is not on $\overline{WX}$. Label it Y.

   d. Draw $\overline{WY}$.

   e. Draw a new point that is not on $\overrightarrow{WY}$ or on the line containing $\overline{WX}$. Label it Z.

   f. Construct $\overleftrightarrow{WZ}$.

   g. Identify $\angle ZWX$ by drawing an arc to indicate the position of the angle.

   h. Identify another angle by referencing points that you have already drawn. _____

Lesson 1:   Identify and draw points, lines, line segments, rays, and angles.
Recognize them in various contexts and familiar figures.

3. a. Observe the familiar figures below. Label some points on each figure.

    b. Use those points to label and name representations of each of the following in the table below: ray, line, line segment, and angle. Extend segments to show lines and rays.

  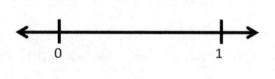

	Clock	Die	Number line
Ray			
Line			
Line segment			
Angle			

Extension: Draw a familiar figure. Label it with points, and then identify rays, lines, line segments, and angles as applicable.

Lesson 1:   Identify and draw points, lines, line segments, rays, and angles.
Recognize them in various contexts and familiar figures.

EUREKA MATH

> I can remake a right angle template using a circle of paper. I fold it into fourths and use the square corner.

1. Use the right angle template that you made in class to determine if each of the following angles is greater than, less than, or equal to a right angle. Label each as *greater than, less than,* or *equal to,* and then connect each angle to the correct label of acute, right, or obtuse.

> I draw a line to "acute" because it names this angle that is less than a right angle.

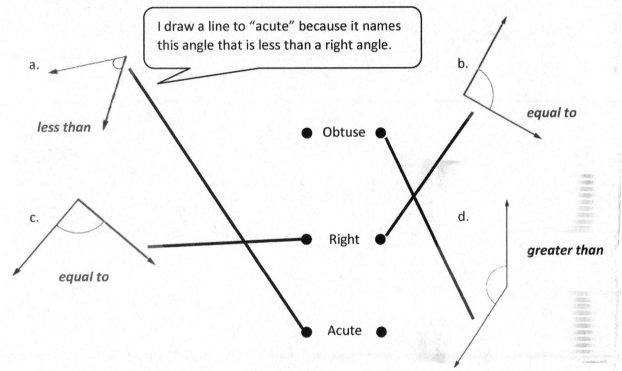

a.

less than

b.

equal to

● Obtuse ●

c.

d.

● Right ●

greater than

equal to

● Acute ●

2. Construct an obtuse angle using a straightedge and the right angle template that you created. Explain the characteristics of an obtuse angle by comparing it to a right angle. Use the words *greater than, less than,* or *equal to* in your explanation.

Sample explanation:

The measure of an obtuse angle is greater than the measure of a right angle.

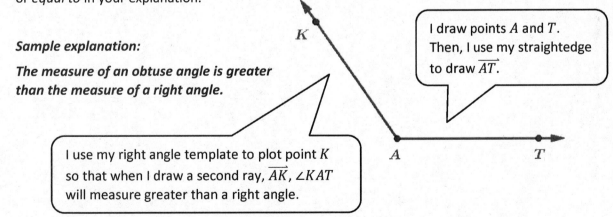

> I draw points A and T. Then, I use my straightedge to draw $\overrightarrow{AT}$.

> I use my right angle template to plot point K so that when I draw a second ray, $\overrightarrow{AK}$, $\angle KAT$ will measure greater than a right angle.

Lesson 2:   Use right angles to determine whether angles are equal to, greater than, or less than right angles. Draw right, obtuse, and acute angles.   311

© 2018 Great Minds®. eureka-math.org

Name _____   Date _____

1. Use the right angle template that you made in class to determine if each of the following angles is greater than, less than, or equal to a right angle. Label each as *greater than*, *less than*, or *equal to,* and then connect each angle to the correct label of acute, right, or obtuse. The first one has been completed for you.

a.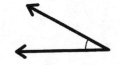

b.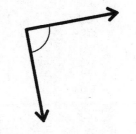

Less than

c.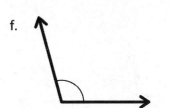

d.

● **Acute** ●

e.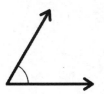

f.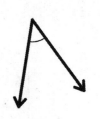

● **Right** ●

● **Obtuse** ●

g.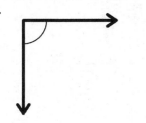

h.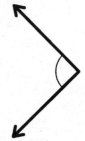

i.

j.

Lesson 2:   Use right angles to determine whether angles are equal to, greater than, or less than right angles. Draw right, obtuse, and acute angles.

© 2018 Great Minds®. eureka-math.org

2. Use your right angle template to identify acute, obtuse, and right angles within this painting.
    Trace at least two of each, label with points, and then name them in the table below the painting.

Acute angle		
Obtuse angle		
Right angle		

Lesson 2: Use right angles to determine whether angles are equal to, greater
than, or less than right angles. Draw right, obtuse, and acute angles.

EUREKA
MATH®

3. Construct each of the following using a straightedge and the right angle template that you created. Explain the characteristics of each by comparing the angle to a right angle. Use the words *greater than, less than,* or *equal to* in your explanations.

   a. Acute angle

   b. Right angle

   c. Obtuse angle

Lesson 2:   Use right angles to determine whether angles are equal to, greater than, or less than right angles. Draw right, obtuse, and acute angles.

© 2018 Great Minds®. eureka-math.org

315

1. On each object, trace at least one pair of lines that appear to be perpendicular.

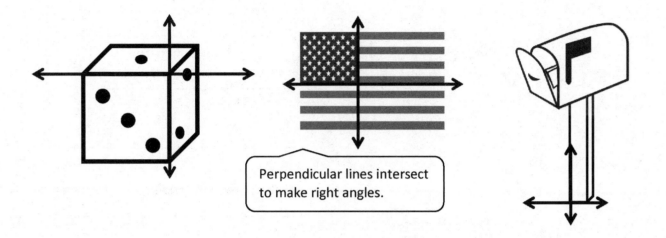

Perpendicular lines intersect to make right angles.

2. In the grid below, draw a segment that is perpendicular to the given segment. Use a straightedge.

I can turn the paper to make the diagonal segment horizontal, if that helps.

The segment perpendicular to this diagonal cuts the triangles in half.

3. Use the right angle template that you created in class to determine if the following figure has a right angle. If so, mark it with a small square. For each right angle you find, name the corresponding perpendicular sides.

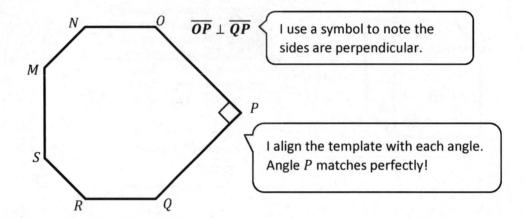

$\overline{OP} \perp \overline{QP}$

I use a symbol to note the sides are perpendicular.

I align the template with each angle. Angle P matches perfectly!

Lesson 3:   Identify, define, and draw perpendicular lines.

EUREKA
MATH

Name _____ Date _____

1. On each object, trace at least one pair of lines that appear to be perpendicular.

2. How do you know if two lines are perpendicular?

3. In the square and triangular grids below, use the given segments in each grid to draw a segment that is perpendicular. Use a straightedge.

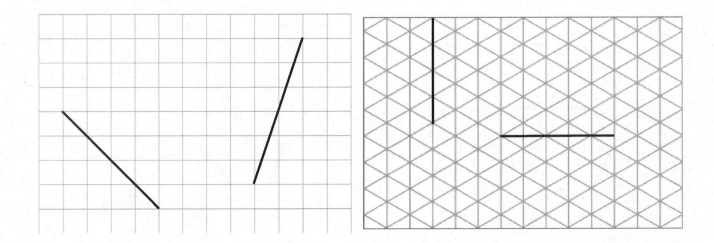

4. Use the right angle template that you created in class to determine which of the following figures have a
    right angle. Mark each right angle with a small square. For each right angle you find, name the
    corresponding pair of perpendicular sides. (Problem 4(a) has been started for you.)

a.

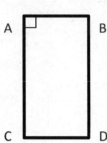

$\overline{CA} \perp \overline{AB}$

b.

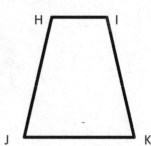

c.

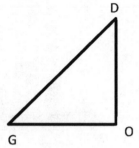

d.

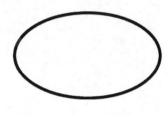

e.

f.

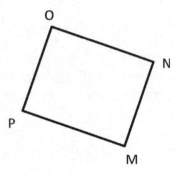

g.

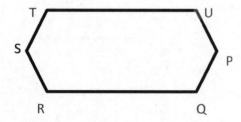

h.

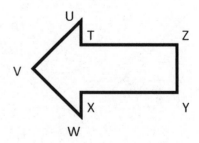

Lesson 3:   Identify, define, and draw perpendicular lines.

EUREKA
MATH

5. Use your right angle template as a guide, and mark each right angle in the following figure with a small square. (Note: A right angle does not have to be inside the figure.) How many pairs of perpendicular sides does this figure have?

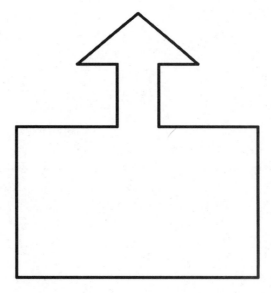

6. True or false? Shapes that have no right angles also have no perpendicular segments. Draw some figures to help explain your thinking.

On each object, trace at least one pair of lines that appear to be parallel.

1.

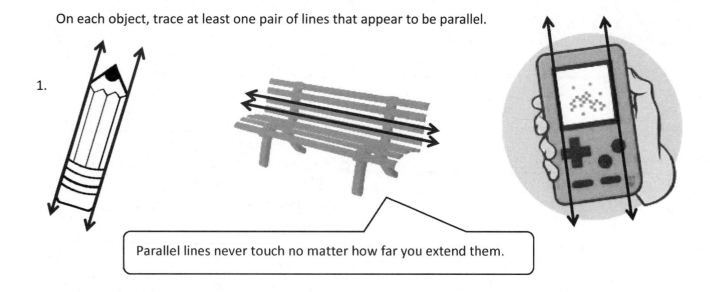

Parallel lines never touch no matter how far you extend them.

In the grid below, use a straightedge to draw a segment that is parallel to the given segment.

2.

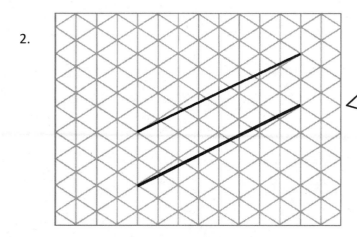

It's tricky to draw diagonal parallel line segments! I draw a line segment that is a distance of two triangle base lengths at every point along the segment.

3. Draw a line using your straightedge. Then, use your right angle template and straightedge to construct a line parallel to the first line you drew.

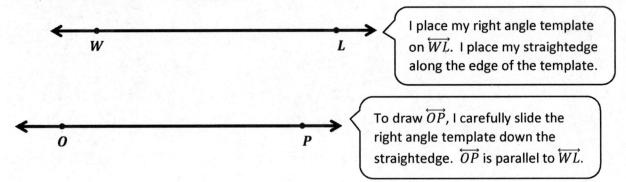

I place my right angle template on $\overleftrightarrow{WL}$. I place my straightedge along the edge of the template.

To draw $\overleftrightarrow{OP}$, I carefully slide the right angle template down the straightedge. $\overleftrightarrow{OP}$ is parallel to $\overleftrightarrow{WL}$.

EUREKA
MATH

Name _____ Date _____

1. On each object, trace at least one pair of lines that appear to be parallel.

2. How do you know if two lines are parallel?

3. In the square and triangular grids below, use the given segments in each grid to draw a segment that is parallel using a straightedge.

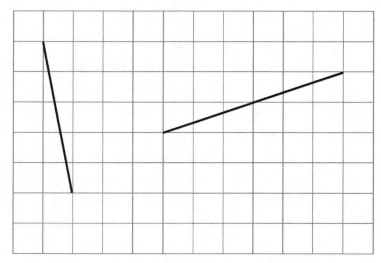 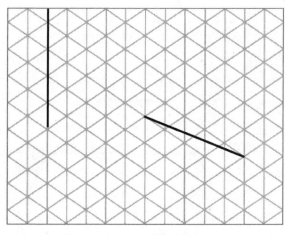

4. Determine which of the following figures have sides that are parallel by using a straightedge and the right angle template that you created. Circle the letter of the shapes that have at least one pair of parallel sides. Mark each pair of parallel sides with arrows, and then identify the parallel sides with a statement modeled after the one in 4(a).

 a.

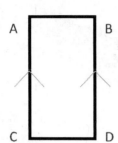

$\overline{AC} \parallel \overline{BD}$

b.

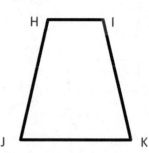

c.

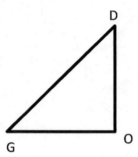

d.

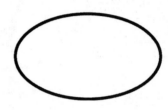

e.

f.

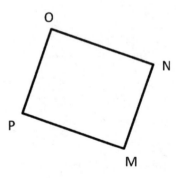

g.

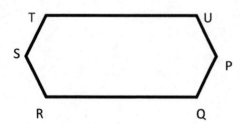

h.

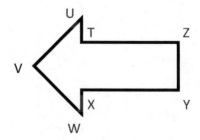

Lesson 4:   Identify, define, and draw parallel lines.

EUREKA
MATH

5. True or false? All shapes with a right angle have sides that are parallel. Explain your thinking.

6. Explain why $\overline{AB}$ and $\overline{CD}$ are parallel, but $\overline{EF}$ and $\overline{GH}$ are not.

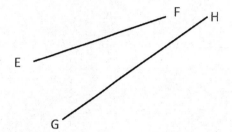

7. Draw a line using your straightedge. Now, use your right angle template and straightedge to construct a line parallel to the first line you drew.

1. Identify the measures of the following angles.

    The angle measures 80°.

    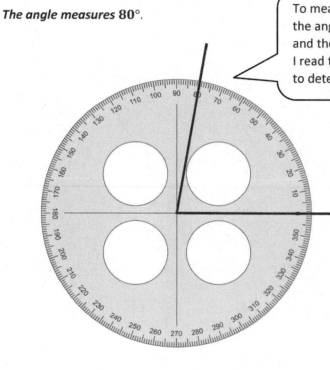

    To measure an angle, I place the protractor on
    the angle so that one of the rays aligns to zero
    and the vertex is at the center of the protractor.
    I read the number aligned with the second ray
    to determine the measure of the angle.

    I use a protractor to measure
    angles. A protractor has tick
    marks like a ruler, but instead
    of measuring inches or
    centimeters, it measures
    degrees around a point.

2. If you didn't have a protractor, how could you construct one? Use words, pictures, or numbers to explain
    in the space below.

    Sample Student Response:

    *If I didn't have a protractor, I could cut out a paper circle. Using a right angle template, I could partition
    the circle in fourths and then mark 0°, 90°, 180°, 270°, and 360°. Although my protractor would not
    be able to give an exact measurement of any angle, I could estimate the measure using these
    benchmarks.*

    I reflect on my experiences and discussions in class.
    We partitioned paper circles in various ways,
    labeling degrees accurately.

EUREKA MATH®   Lesson 5:   Use a circular protractor to understand a 1-degree angle as $\frac{1}{360}$ of a   **329**
                                    turn. Explore benchmark angles using the protractor.

© 2018 Great Minds®. eureka-math.org

Name _____ Date _____

1. Identify the measures of the following angles.

a.

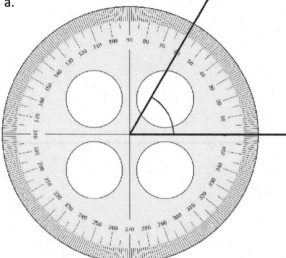

b.

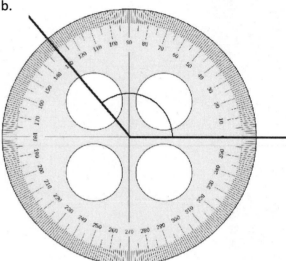

c.

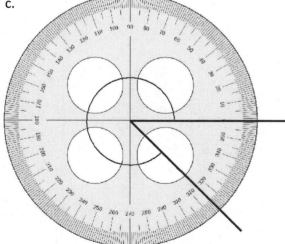

d.

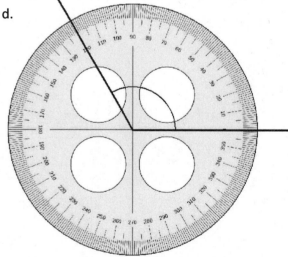

EUREKA MATH

Lesson 5: Use a circular protractor to understand a 1-degree angle as $\frac{1}{360}$ of a turn. Explore benchmark angles using the protractor.

331

© 2018 Great Minds®. eureka-math.org

2. If you didn't have a protractor, how could you construct one? Use words, pictures, or numbers to explain in the space below.

Lesson 5:   Use a circular protractor to understand a 1-degree angle as $\frac{1}{360}$ of a turn. Explore benchmark angles using the protractor.

EUREKA
MATH

1. Use a protractor to measure the angle, and then record the measurement in degrees.

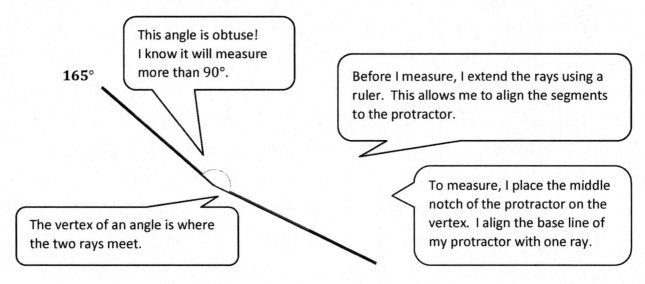

165°

This angle is obtuse! I know it will measure more than 90°.

Before I measure, I extend the rays using a ruler. This allows me to align the segments to the protractor.

The vertex of an angle is where the two rays meet.

To measure, I place the middle notch of the protractor on the vertex. I align the base line of my protractor with one ray.

2. Use a protractor to measure the angle. Extend the length of the segments as needed. When you extend the segments, does the angle measure stay the same? Explain how you know.

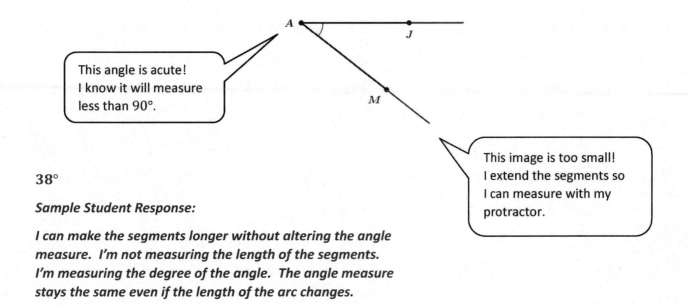

This angle is acute! I know it will measure less than 90°.

This image is too small! I extend the segments so I can measure with my protractor.

38°

Sample Student Response:

I can make the segments longer without altering the angle measure. I'm not measuring the length of the segments. I'm measuring the degree of the angle. The angle measure stays the same even if the length of the arc changes.

EUREKA MATH

Lesson 6:   Use varied protractors to distinguish angle measure from length measurement.

© 2018 Great Minds®. eureka-math.org

333

Name _____   Date _____

1. Use a protractor to measure the angles, and then record the measurements in degrees.

a.

b.

c.

d.

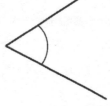

EUREKA MATH®

Lesson 6:   Use varied protractors to distinguish angle measure from length measurement.

335

© 2018 Great Minds®. eureka-math.org

e.

f.

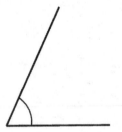

g.

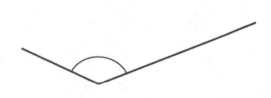

h.

i.

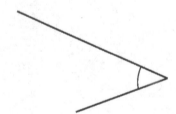

j.

Lesson 6: Use varied protractors to distinguish angle measure from length measurement.

EUREKA
MATH

2. Using the green and red circle cutouts from today's lesson, explain to someone at home how the cutouts can be used to show that the angle measures are the same even though the circles are different sizes. Write words to explain what you told him or her.

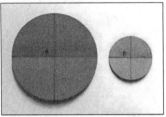

3. Use a protractor to measure each angle. Extend the length of the segments as needed. When you extend the segments, does the angle measure stay the same? Explain how you know.

a.

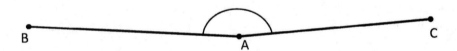

b.

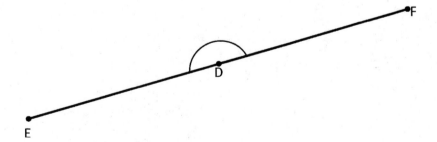

EUREKA
MATH

Lesson 6:   Use varied protractors to distinguish angle measure from length
               measurement.

337

© 2018 Great Minds®. eureka-math.org

Construct angles that measure the give number of degrees. For the first problem, use the ray shown as one of the rays of the angle with its endpoint as the vertex of the angle. Draw an arc to indicate the angle that was measured.

1. 90°

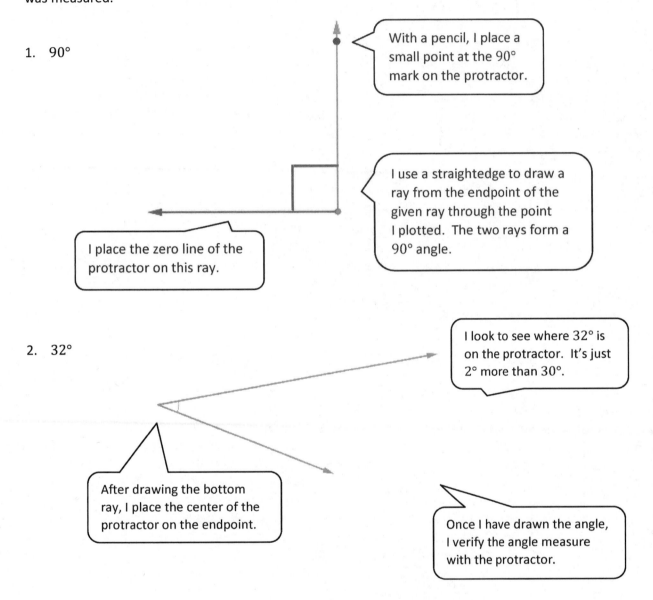

With a pencil, I place a small point at the 90° mark on the protractor.

I use a straightedge to draw a ray from the endpoint of the given ray through the point I plotted. The two rays form a 90° angle.

I place the zero line of the protractor on this ray.

2. 32°

I look to see where 32° is on the protractor. It's just 2° more than 30°.

After drawing the bottom ray, I place the center of the protractor on the endpoint.

Once I have drawn the angle, I verify the angle measure with the protractor.

EUREKA MATH

Lesson 7:   Measure and draw angles. Sketch given angle measures, and verify with a protractor.   **339**

© 2018 Great Minds®. eureka-math.org

Name _____   Date _____

Construct angles that measure the given number of degrees. For Problems 1–4, use the ray shown as one of the rays of the angle with its endpoint as the vertex of the angle. Draw an arc to indicate the angle that was measured.

1.   25°

2.   85°

3.   140°

4.   83°

Lesson 7:   Measure and draw angles. Sketch given angle measures, and verify with a protractor.

341

EUREKA
MATH

5. 108°

6. 72°

7. 25°

8. 155°

9. 45°

10. 135°

Lesson 7:   Measure and draw angles. Sketch given angle measures, and verify
            with a protractor.

© 2018 Great Minds®. eureka-math.org

EUREKA
MATH

1. James looked at the clock when he put the cake in the oven and when he took it out. How many degrees did the minute hand turn from start to finish?

start time

end time

The minute hand turned 180°.

I know from Lesson 5 that there are 360° in a full turn. From the 12 to the 3 is a 90° angle, and from the 3 to the 6 is another 90° angle.

2. Delonte turned the lock on his locker one quarter turn to the right and then 180° to the left. Draw a picture showing the position of the lock after he turned it.

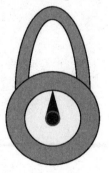

before

after

I think of the lock as a clock. A quarter-turn to the right is 15 minutes, and 180° to the left is 30 minutes backward.

3. How many quarter-turns does the picture need to be rotated in order for it to be upright?

To be upright, the picture needs to be turned two quarter-turns.

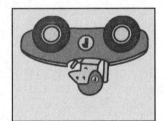

I can turn the paper itself to help me figure out the answer!

Lesson 8: Identify and measure angles as turns and recognize them in various contexts.

© 2018 Great Minds®. eureka-math.org

343

Name _____   Date _____

1. Jill, Shyan, and Barb stood in the middle of the yard and faced the barn. Jill turned 90° to the right.
    Shyan turned 180° to the left. Barb turned 270° to the left. Name the object that each girl is now facing.

    Jill _____

    Shyan _____

    Barb _____

    House

    Barn   Fence

          Yard

          Tree

2. Allison looked at the clock at the beginning of class and at the end of class. How many degrees did the
    minute hand turn from the beginning of class until the end?

    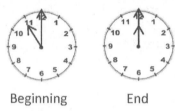

    Beginning   End

3. The snowboarder went off a jump and did a 180. In which direction was the snowboarder facing when he
    landed? How do you know?

4. As she drove down the icy road, Mrs. Campbell slammed on her brakes. Her car did a 360. Explain what
    happened to Mrs. Campbell's car.

EUREKA
MATH

Lesson 8:   Identify and measure angles as turns and recognize them in various
             contexts.

© 2018 Great Minds®. eureka-math.org

345

5. Jonah turned the knob of the stove two quarter-turns. Draw a picture showing the position of the knob after he turned it.

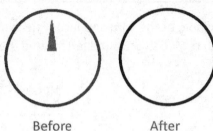

Before   After

6. Betsy used her scissors to cut out a coupon from the newspaper. How many total quarter-turns will she need to rotate the paper in order to cut out the entire coupon?

7. How many quarter-turns does the picture need to be rotated in order for it to be upright?

8. David faced north. He turned 180° to the right, and then 270° to the left. In which direction is he now facing?

Lesson 8:   Identify and measure angles as turns and recognize them in various contexts.

© 2018 Great Minds®. eureka-math.org

EUREKA
MATH®

Pattern Blocks

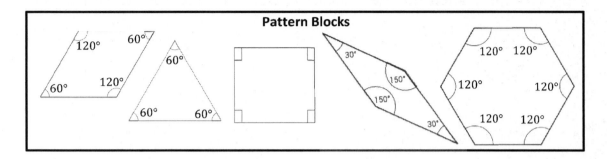

Sketch one way to compose ∠ABC using two or more pattern blocks. Write an addition sentence to show how you composed the given angle.

1. ∠ABC = 150°

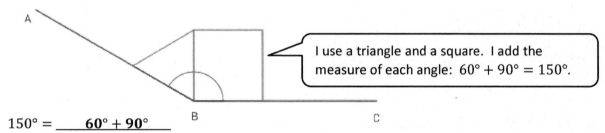

I use a triangle and a square. I add the measure of each angle: $60° + 90° = 150°$.

150° = ____ **60° + 90°** ____

Sabrina built the following shape with her pattern blocks. As indicated by their arcs, solve for x°, y°, and z°. Write an addition sentence for each. The first one is done for you.

2.

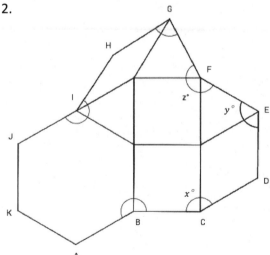

a. $y° = 60° + 60°$

$y° = 120°$

b. $z° =$ ____ **60° + 90° + 60°** ____

$z° =$ ____ **210°** ____

c. $x° =$ ____ **90° + 60°** ____

$x° =$ ____ **150°** ____

To determine x°, y°, and z°, I add together the smaller angles encompassed by the arcs. I use the chart at the top of the page to determine the measure of each of the smaller angles.

Name _____   Date _____

Sketch two different ways to compose the given angles using two or more pattern blocks.
Write an addition sentence to show how you composed the given angle.

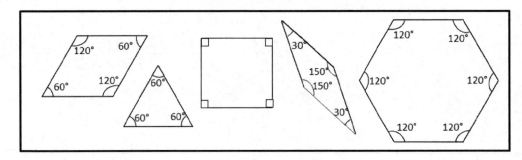

1. Points *A*, *B*, and *C* form a straight line.

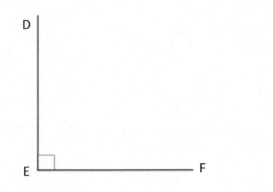

A   B   C   A   B   C

180° = _____   180° = _____

2. ∠*DEF* = 90°

D   D

E   F   E   F

90° = _____   90° = _____

3. $\angle GHI = 120°$

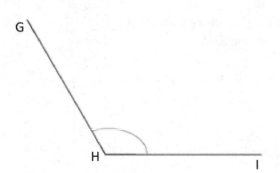

$120° = $ _____

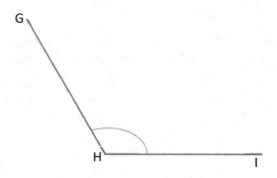

$120° = $ _____

4. $x° = 270°$

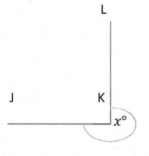

$270° = $ _____

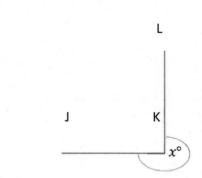

$270° = $ _____

5. Micah built the following shape with his pattern blocks. Write an addition sentence for each angle indicated by an arc and solve. The first one is done for you.

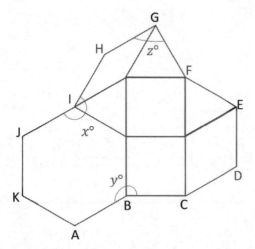

a. $y° = 120° + 90°$

$y° = 210°$

b. $z° = $ _____

$z° = $ _____

c. $x° = $ _____

$x° = $ _____

Lesson 9:   Decompose angles using pattern blocks.

EUREKA
MATH

1. Write an equation, and solve for the measurement of $\angle x$. Verify the measurement using a protractor.

a. $\angle JKL$ is a straight angle.

b. Solve for the measurement of $\angle USW$. $\angle RST$ is a straight angle.

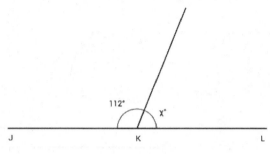

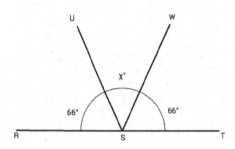

$$\underline{112°} + \underline{68°} = \underline{180°}$$

$$x° = \underline{68°}$$

$$\begin{array}{r} 7 \quad 10 \\ 1 \quad \cancel{8} \quad \cancel{0} \\ -\ 1 \quad 3 \quad 2 \\ \hline 0 \quad 4 \quad 8 \end{array}$$

$$66° + 66° + x° = 180°$$
$$132° + x° = 180°$$
$$x° = 48°$$
$$\angle USW = 48°$$

I know a straight angle measures 180°.
I subtract 112° from 180° to find the value of $x°$.
To verify my answer, I use my protractor to measure the angle. It measures 68°.

I know that the sum of these three angle measures is 180°. I add the two parts that I know and then I subtract their total from 180°.

2. Complete the following directions in the space to the right.

a. Draw 2 points: S and T. Using a straightedge, draw $\overleftrightarrow{ST}$.

b. Plot a point U somewhere between points S and T.

c. Plot a point W, which is not on $\overleftrightarrow{ST}$.

d. Draw $\overline{UW}$.

e. Find the measure of $\angle SUW$ and $\angle TUW$.

f. Write an equation to show that the angles add to the measure of a straight angle.

Sample Response:

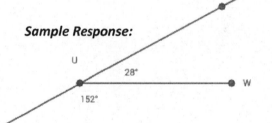

I draw the figure. I use my protractor to measure $\angle SUW$ and $\angle TUW$.

$$\angle SUW = 152°$$

$$\angle TUW = 28°$$

$$152° + 28° = 180°$$

Lesson 10:   Use the addition of adjacent angle measures to solve problems using a symbol for the unknown angle measure.

351

© 2018 Great Minds®. eureka-math.org

Name _____   Date _____

Write an equation, and solve for the measurement of ∠x. Verify the measurement using a protractor.

1. ∠DCB is a right angle.

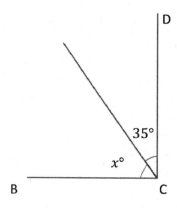

_____ + 35° = 90°

$x° =$ _____

2. ∠HGF is a right angle.

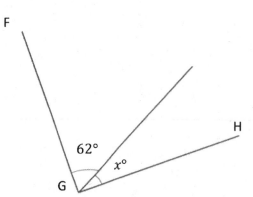

_____ + _____ = _____

$x° =$ _____

3. ∠JKL is a straight angle.

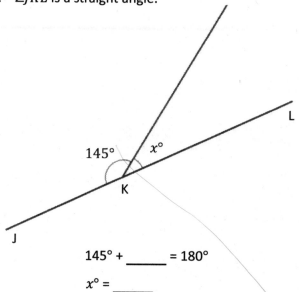

145° + _____ = 180°

$x° =$ _____

4. ∠PQR is a straight angle.

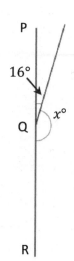

_____ + _____ = _____

$x° =$ _____

EUREKA
MATH

Lesson 10:   Use the addition of adjacent angle measures to solve problems using a
symbol for the unknown angle measure.

© 2018 Great Minds®. eureka-math.org

353

Write an equation, and solve for the unknown angle measurements.

5. Solve for the measurement of ∠USW.
    ∠RST is a straight angle.

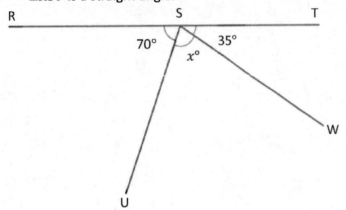

6. Solve for the measurement of ∠OML.
    ∠LMN is a straight angle.

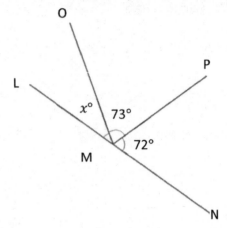

7. In the following figure, DEFH is a rectangle. Without using a protractor, determine the measurement of ∠GEF. Write an equation that could be used to solve the problem.

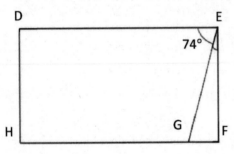

8. Complete the following directions in the space to the right.

    a. Draw 2 points: Q and R. Using a straightedge, draw $\overleftrightarrow{QR}$.

    b. Plot a point S somewhere between points Q and R.

    c. Plot a point T, which is not on $\overleftrightarrow{QR}$.

    d. Draw $\overline{TS}$.

    e. Find the measure of ∠QST and ∠RST.

    f. Write an equation to show that the angles add to the measure of a straight angle.

Use the addition of adjacent angle measures to solve problems using a symbol for the unknown angle measure.

© 2018 Great Minds®. eureka-math.org

Write an equation, and solve for the unknown angle measurements numerically.

1.

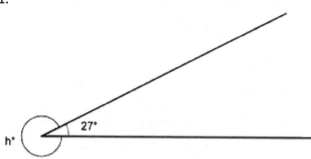

> I know from Lesson 5 that a circle measures 360°.
> I solve for $h°$ by subtracting 27° from 360°.

$$\underline{\ 27°\ } + \underline{\ 333°\ } = 360°$$

$$h° = \underline{\ 333°\ }$$

$$\begin{array}{r} {\scriptstyle 5 \quad 10} \\ 3\ \ \cancel{6}\ \ \cancel{0} \\ -\ \ \ \ 2\ \ 7 \\ \hline 3\ \ 3\ \ 3 \end{array}$$

2.

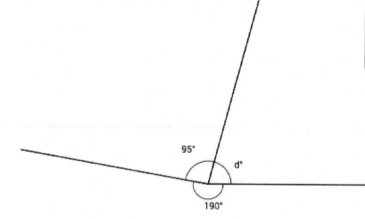

> I solve for $d°$ by adding together the known angle measures and then subtracting their sum from 360°.

$$\underline{\ 190°\ } + \underline{\ 95°\ } + \underline{\ 75°\ } = \underline{\ 360°\ }$$

$$d° = \underline{\ 75°\ }$$

$$\begin{array}{r} 1\ \ 9\ \ 0 \\ +\ \ \ \ 9\ \ 5 \\ \hline {\scriptstyle 1} \\ 2\ \ 8\ \ 5 \end{array} \qquad \begin{array}{r} {\scriptstyle \quad\ \ 15} \\ {\scriptstyle 2\ \ \cancel{5}\ \ 10} \\ \cancel{3}\ \ \cancel{6}\ \ \cancel{0} \\ -\ 2\ \ 8\ \ 5 \\ \hline 7\ \ 5 \end{array}$$

EUREKA MATH®   **Lesson 11:**   Use the addition of adjacent angle measures to solve problems using a   **355**
                               symbol for the unknown angle measure.

© 2018 Great Minds®. eureka-math.org

3. T is the intersection of $\overline{UV}$ and $\overline{WX}$.
   $\angle UTW$ is 51°.

$g° =$ __129°__ $h° =$ __51°__ $i° =$ __129°__

$$129° + h° = 180° \qquad 51° + i° = 180°$$
$$h° = 51° \qquad\qquad i° = 129°$$

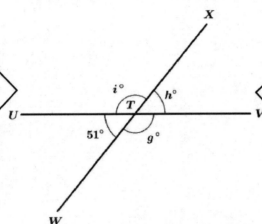

I can solve for $i°$ by thinking of its relationship to either $\overline{UV}$ or $\overline{WX}$. But I also notice that opposite angles measure the same for this figure.

I solve for $h°$ by thinking about the relationships of $\angle WTV$ and $\angle VTX$. Both angle measures add to 180° because they are on $\overline{WX}$.

$$51° + g° = 180°$$
$$g° = 129°$$

I solve for $g°$ by thinking of its relationship to $\angle UTW$. $\angle UTV$ is a straight angle that measures 180°.

$$
\begin{array}{r}
{\scriptstyle 7 \ \ 10} \\
1 \ \cancel{8} \ \cancel{0} \\
- \quad 5 \ \ 1 \\
\hline
1 \ \ 2 \ \ 9
\end{array}
$$

Lesson 11:   Use the addition of adjacent angle measures to solve problems using a symbol for the unknown angle measure.

© 2018 Great Minds®. eureka-math.org

EUREKA MATH

4. P is the intersection of $\overline{QR}$, $\overline{ST}$, and $\overline{UP}$. $j° =$ ___**124°**___ $k° =$ ___**56°**___ $m° =$ ___**34°**___
    $\angle QPS$ is 56°.

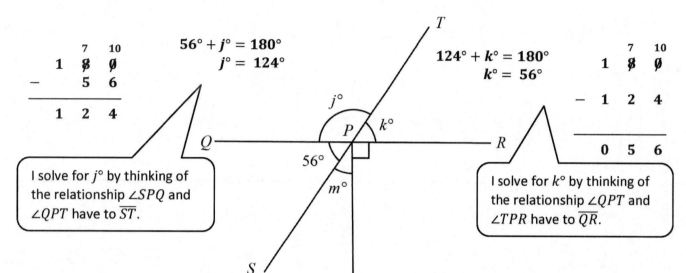

$$\begin{array}{r} 7 \quad 10 \\ 1 \; \cancel{8} \; \cancel{0} \\ - \quad 5 \quad 6 \\ \hline 1 \quad 2 \quad 4 \end{array}$$

$$56° + j° = 180°$$
$$j° = 124°$$

$$124° + k° = 180°$$
$$k° = 56°$$

$$\begin{array}{r} 7 \quad 10 \\ 1 \; \cancel{8} \; \cancel{0} \\ - \quad 1 \quad 2 \quad 4 \\ \hline 0 \quad 5 \quad 6 \end{array}$$

I solve for $j°$ by thinking of the relationship $\angle SPQ$ and $\angle QPT$ have to $\overline{ST}$.

I solve for $k°$ by thinking of the relationship $\angle QPT$ and $\angle TPR$ have to $\overline{QR}$.

I solve for $m°$ by noticing that $\angle UPR$ is a right angle; therefore, $\angle UPQ$ is also a right angle.

$$56° + m° = 90°$$
$$m° = 34°$$

$$\begin{array}{r} 8 \quad 10 \\ \cancel{9} \; \cancel{0} \\ - \quad 5 \quad 6 \\ \hline 3 \quad 4 \end{array}$$

EUREKA
MATH

© 2018 Great Minds®. eureka-math.org

Name _____   Date _____

Write an equation, and solve for the unknown angle measurements numerically.

1.

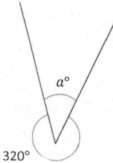

320°

_____° + 320° = 360°

$a° =$ _____°

2.

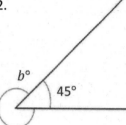

_____° + _____° = 360°

$b° =$ _____°

3.

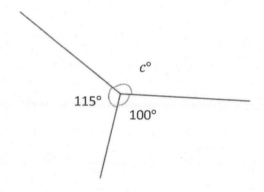

_____° + _____° + _____° = _____°

$c° =$ _____°

4.

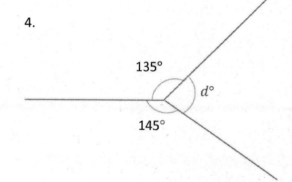

_____° + _____° + _____° = _____°

$d° =$ _____°

EUREKA
MATH

Lesson 11:   Use the addition of adjacent angle measures to solve problems using a
symbol for the unknown angle measure.

© 2018 Great Minds®. eureka-math.org

359

Write an equation, and solve for the unknown angles numerically.

5. O is the intersection of $\overline{AB}$ and $\overline{CD}$.
   $\angle COB$ is 145°, and $\angle AOC$ is 35°.

   $e° =$ _____ $f° =$ _____

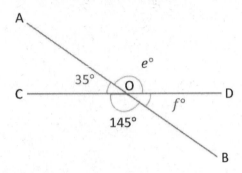

6. O is the intersection of $\overline{QR}$ and $\overline{ST}$.
   $\angle QOS$ is 55°.

   $g° =$ _____ $h° =$ _____ $i° =$ _____

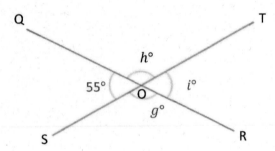

7. O is the intersection of $\overline{UP}$, $\overline{WX}$, and $\overline{YO}$.
   $\angle VOX$ is 46°.

   $j° =$ _____ $k° =$ _____ $m° =$ _____

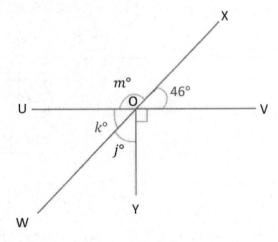

Lesson 11:   Use the addition of adjacent angle measures to solve problems using a symbol for the unknown angle measure.

EUREKA
MATH

I can tell parts (b) and (d) each have a line of symmetry because the figure in each part is the same on both sides of the line.

1. Circle the figures that have a correct line of symmetry drawn.

a. 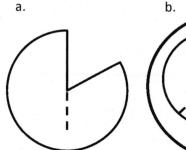   b. 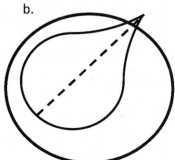   c.   d.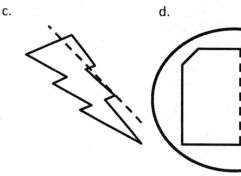

2. Find and draw all lines of symmetry for the following figures. Write the number of lines of symmetry that you found in the blank underneath the shape.

   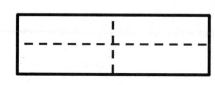   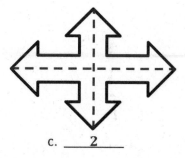

a. __1__   b. __2__   c. __2__

I think about folding these shapes in half many different ways. If the shapes match where I fold them, that is a line of symmetry.

EUREKA MATH

Lesson 12:   Recognize lines of symmetry for given two-dimensional figures. Identify line-symmetric figures, and draw lines of symmetry.

© 2018 Great Minds®. eureka-math.org

361

3. Half of the figure below has been drawn. Use the line of symmetry, represented by the dashed line, to complete the figure.

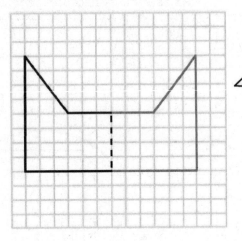

I use the grid to help me complete the figure. I count how many units long each segment is, and then I draw segments of the same length for the other half of the figure. I draw the sides that follow the grid lines first, and then I make the diagonal line.

Lesson 12: Recognize lines of symmetry for given two-dimensional figures. Identify line-symmetric figures, and draw lines of symmetry.

© 2018 Great Minds®. eureka-math.org

EUREKA MATH

Name _____ Date _____

1. Circle the figures that have a correct line of symmetry drawn.

a.    b.    c.   d.

2. Find and draw all lines of symmetry for the following figures. Write the number of lines of symmetry that you found in the blank underneath the shape.

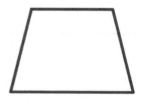

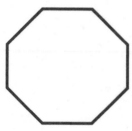

a. _____   b. _____   c. _____

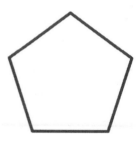

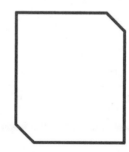

d. _____   e. _____   f. _____

g. _____   h. _____   i. _____

EUREKA MATH

Lesson 12: Recognize lines of symmetry for given two-dimensional figures.
Identify line-symmetric figures, and draw lines of symmetry.

363

© 2018 Great Minds®. eureka-math.org

3. Half of each figure below has been drawn. Use the line of symmetry, represented by the dashed line, to complete each figure.

a.

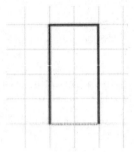

b.

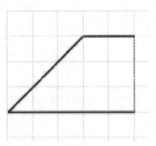

c.

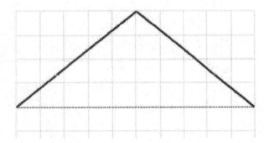

d.

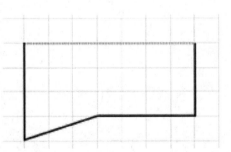

4. Is there another shape that has the same number of lines of symmetry as a circle? Explain.

Lesson 12: Recognize lines of symmetry for given two-dimensional figures. Identify line-symmetric figures, and draw lines of symmetry.

EUREKA MATH

1. Classify each triangle by its side lengths and angle measurements. Circle the correct names.

	Classify Using Side Lengths	Classify Using Angle Measurements
a.	Equilateral Isosceles (Scalene)	Acute (Right) Obtuse
b.	Equilateral (Isosceles) Scalene	Acute Right (Obtuse)
c.	(Equilateral) Isosceles Scalene	(Acute) Right Obtuse

Sometimes triangles are drawn with tick marks, little dashes perpendicular to the sides of the triangle. These tick marks mean that those sides have the same length.

To classify by side lengths, I use a ruler to measure each side of the triangle or look to see if tick marks are drawn. Equilateral triangles have sides that are all the same length. Isosceles triangles have two sides that are the same length. Scalene triangles have sides that are all different lengths.

To classify by angle measure, I can use a protractor or a right angle template. An acute triangle has three angles less than 90°.

A right triangle has one 90° angle. An obtuse triangle has one angle greater than 90°.

EUREKA
MATH

Lesson 13:   Analyze and classify triangles based on side length, angle measure, or both.

© 2018 Great Minds®. eureka-math.org

365

2. Use a ruler to connect points to form two other triangles. Use each point only once. None of the triangles may overlap. One point will be unused. Name and classify the three triangles below. The first one has been done for you.

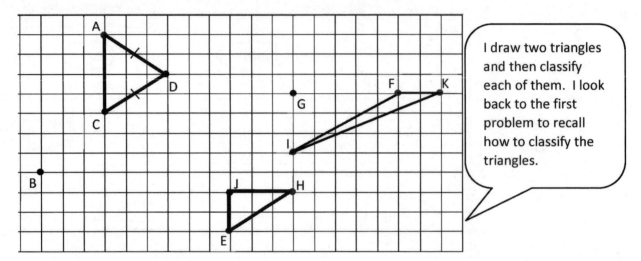

I draw two triangles and then classify each of them. I look back to the first problem to recall how to classify the triangles.

Name the Triangles Using Vertices	Classify by Side Length	Classify by Angle Measurement
△ FKI	Scalene	Obtuse
△ ACD	Isosceles	Acute
△ EHJ	Scalene	Right

3. Can a triangle have two obtuse angles? Explain.

Sample answer:

No, if a triangle had two obtuse angles, the three sides could never meet.

I draw two obtuse angles, and I see that the three sides can't form a triangle since two of the line segments will continue to get farther apart instead of closer together if I make them longer.

Lesson 13:   Analyze and classify triangles based on side length, angle measure, or both.

© 2018 Great Minds®. eureka-math.org

Name _____ Date _____

1. Classify each triangle by its side lengths and angle measurements. Circle the correct names.

	Classify Using Side Lengths	Classify Using Angle Measurements
a.	Equilateral Isosceles Scalene	Acute Right Obtuse
b.	Equilateral Isosceles Scalene	Acute Right Obtuse
c.	Equilateral Isosceles Scalene	Acute Right Obtuse
d.	Equilateral Isosceles Scalene	Acute Right Obtuse

2. a. △ ABC has one line of symmetry as shown. Is the measure of ∠A greater than, less than, or equal to ∠C?

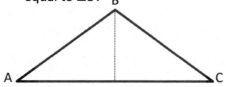

b. △ DEF is scalene. What do you observe about its angles? Explain.

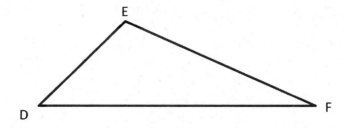

EUREKA
MATH®

Lesson 13:   Analyze and classify triangles based on side length, angle measure, or both.

367

© 2018 Great Minds®. eureka-math.org

3. Use a ruler to connect points to form two other triangles. Use each point only once. None of the triangles may overlap. Two points will be unused. Name and classify the three triangles below.

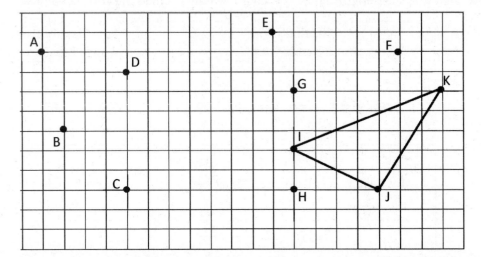

Name the Triangles Using Vertices	Classify by Side Length	Classify by Angle Measurement
△ IJK		

4. If the perimeter of an equilateral triangle is 15 cm, what is the length of each side?

5. Can a triangle have more than one obtuse angle? Explain.

6. Can a triangle have one obtuse angle and one right angle? Explain.

Lesson 13: Analyze and classify triangles based on side length, angle measure, or both.

© 2018 Great Minds®. eureka-math.org

EUREKA MATH

1. Draw triangles that fit the following classifications. Use a ruler and protractor. Label the side lengths and angles.

   a. Acute and equilateral

   b. Right and isosceles

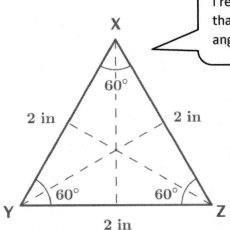

I remember from Lesson 9 that an equilateral triangle has angle measurements of 60°.

To draw this triangle, I first use my protractor to draw the right angle. Then I use my ruler to make sure $\overline{EG}$ and $\overline{GF}$ are the same length.

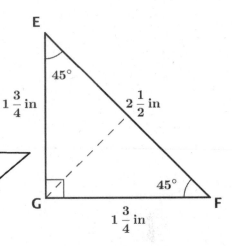

2. Draw all possible lines of symmetry in the triangles above.

   △ XYZ has three lines of symmetry because it is an equilateral triangle.
   △ EFG has one line of symmetry because it is an isosceles triangle.

3. △ EFG can be described as a right triangle and a scalene triangle. True or False?

   Sample answer:

   False. △ EFG *is isosceles and right. I know this because two of the sides are the same length, and there is a right angle.*

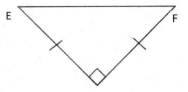

EUREKA
MATH®

Lesson 14: Define and construct triangles from given criteria. Explore symmetry in triangles.

369

© 2018 Great Minds®. eureka-math.org

4. If △ *ABC* is an equilateral triangle, $\overline{BC}$ must be 1 cm. True or False?

   Sample answer:

   True. If △ *ABC* is equilateral, that means that all of the side lengths must be the same length. So, if two of the sides are 1 cm, the third side must also be 1 cm.

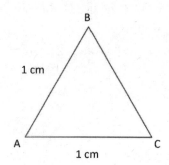

     Lesson 14:   Define and construct triangles from given criteria. Explore symmetry in triangles.

EUREKA
MATH

Name _____   Date _____

1. Draw triangles that fit the following classifications. Use a ruler and protractor. Label the side lengths and angles.

    a.   Right and isosceles

    b.   Right and scalene

    c.   Obtuse and isosceles

    d.   Acute and scalene

2. Draw all possible lines of symmetry in the triangles above. Explain why some of the triangles do not have lines of symmetry.

Lesson 14:   Define and construct triangles from given criteria. Explore symmetry in triangles.

© 2018 Great Minds®. eureka-math.org

371

Are the following statements true or false? Explain.

3. △ **ABC** is an isosceles triangle. $\overline{AB}$ must be 2 cm. True or False?

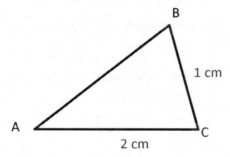

4. A triangle cannot have both an acute angle and a right angle. True or False?

5. △ **XYZ** can be described as both equilateral and acute. True or False?

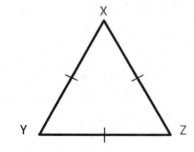

6. A right triangle is always scalene. True or False?

Extension: In △ **ABC**, x = y. True or False?

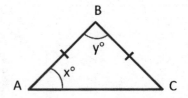

Lesson 14:   Define and construct triangles from given criteria. Explore symmetry
in triangles.

© 2018 Great Minds®. eureka-math.org

EUREKA
MATH

I use what I learned in Lessons 3 and 4 to draw parallel and perpendicular lines using a right angle template and a ruler.

Construct the following figures based on the given attributes. Give a name to each figure you construct. Be as specific as possible.

1. A quadrilateral with opposite sides the same length and four right angles

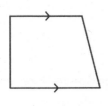

rectangle

I draw the bottom segment using my ruler. I draw the two sides using my right angle template and ruler to make right angles and to make the left and right side lengths equal. I draw the top segment perpendicular to the sides and parallel to the bottom segment. I draw small squares to show the right angles and tick marks to show which sides are equal.

2. A quadrilateral with one set of parallel sides

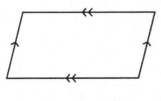

trapezoid

I draw a horizontal segment. I draw a segment that is parallel to the first segment. I connect the endpoints of the segments. I draw arrows to label the parallel sides.

3. A quadrilateral with two sets of parallel sides

parallelogram

I start by drawing horizontal, parallel sides just as when I started drawing a trapezoid. After I draw the left side segment, I make sure the right side segment is parallel to it. I add arrows on the opposite segments to show they are parallel to each other.

Lesson 15:   Classify quadrilaterals based on parallel and perpendicular lines and
             the presence or absence of angles of a specified size.

373

© 2018 Great Minds®. eureka-math.org

4. A parallelogram with all sides the same length and four right angles

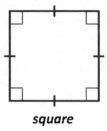

square

> I start by drawing a parallelogram, except I draw the left side segment perpendicular to the horizontal segments. I measure the left side segment and make sure to make the top and bottom segments the same lengths. I draw a right segment perpendicular to the top and bottom segments. It will be the same length as all other sides. I add tick marks and right angle squares.

Lesson 15: Classify quadrilaterals based on parallel and perpendicular lines and the presence or absence of angles of a specified size.

EUREKA MATH

Name _____ Date _____

1. Use the word bank to name each shape, being as specific as possible.

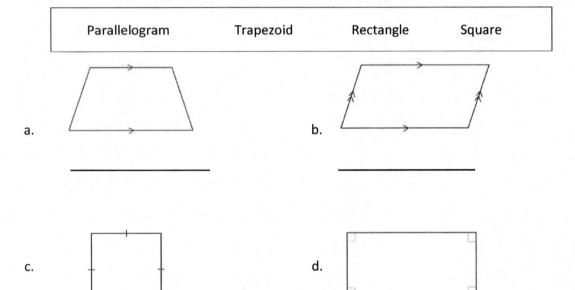

| Parallelogram | Trapezoid | Rectangle | Square |

a. _____

b. _____

c. _____

d. _____

2. Explain the attribute that makes a square a special rectangle.

3. Explain the attribute that makes a rectangle a special parallelogram.

4. Explain the attribute that makes a parallelogram a special trapezoid.

EUREKA
MATH

Lesson 15: Classify quadrilaterals based on parallel and perpendicular lines and the presence or absence of angles of a specified size.

© 2018 Great Minds®. eureka-math.org

375

5. Construct the following figures based on the given attributes. Give a name to each figure you construct. Be as specific as possible.

    a. A quadrilateral with four sides the same length and four right angles.

    b. A quadrilateral with two sets of parallel sides.

    c. A quadrilateral with only one set of parallel sides.

    d. A parallelogram with four right angles.

Lesson 15: Classify quadrilaterals based on parallel and perpendicular lines and the presence or absence of angles of a specified size.

© 2018 Great Minds®. eureka-math.org

EUREKA MATH

1. Construct a quadrilateral with all sides of equal length. What shape did you create?

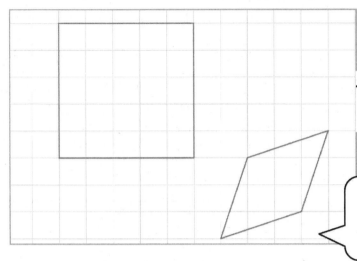

Sample Response:

I created a square.

I trace the gridlines to draw line segments of equal length, constructing a square.

I created a rhombus.

I look for a pattern on the grid to draw a rhombus. I draw segments that go diagonally across three squares of the grid.

2. Construct a quadrilateral with two sets of parallel sides. What shape did you create?

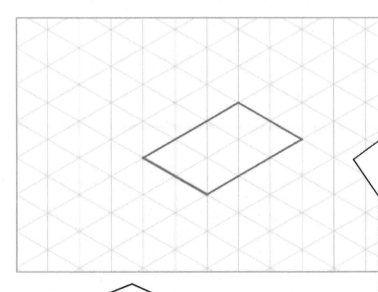

Sample Response:

I created a parallelogram.

I trace along one of the diagonal gridlines. I draw a second segment parallel to the first by tracing along a gridline two triangle side lengths away. I draw the third and fourth segments by tracing along two other diagonal gridlines going in the opposite direction. I use a ruler and right angle template to verify that the sets of sides are parallel.

I also could have drawn a rectangle, a square, or a rhombus because they are also

Lesson 16:   Reason about attributes to construct quadrilaterals on square or triangular grid paper.

Name _____ Date _____

Use the grid to construct the following. Name the figure you drew using one of the terms in the word box.

1. Construct a quadrilateral with only one set of parallel sides.
   Which shape did you create?

WORD BOX
WORD BOX
Parallelogram
Trapezoid
Rectangle
Square
Rhombus

2. Construct a quadrilateral with one set of parallel sides and two right angles.
   Which shape did you create?

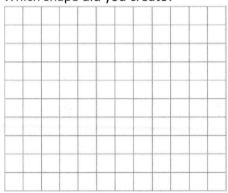

3. Construct a quadrilateral with two sets of parallel sides.
   Which shape did you create?

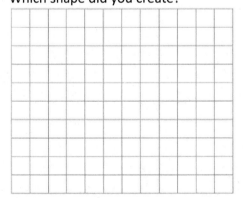

EUREKA MATH

Lesson 16:   Reason about attributes to construct quadrilaterals on square or triangular grid paper.

379

4. Construct a quadrilateral with all sides of equal length.
    Which shape did you create?

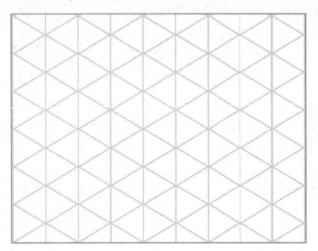

5. Construct a rectangle with all sides of equal length.
    Which shape did you create?

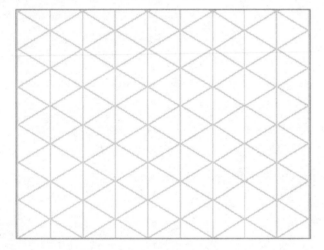

Lesson 16:   Reason about attributes to construct quadrilaterals on square or
                             triangular grid paper.

EUREKA
MATH

Credits

Great Minds® has made every effort to obtain permission for the reprinting of all copyrighted material. If any owner of copyrighted material is not acknowledged herein, please contact Great Minds for proper acknowledgment in all future editions and reprints of this module.